Where the Grass Still Sings

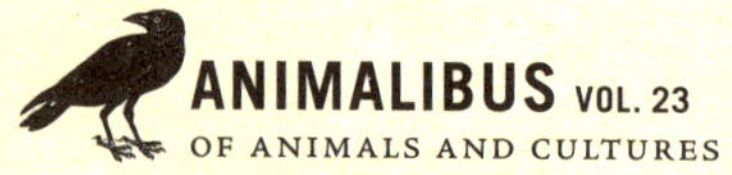

ANIMALIBUS VOL. 23
OF ANIMALS AND CULTURES

Books in the Animalibus series share a fascination with the status and the role of animals in human life. Crossing the humanities and the social sciences to include work in history, anthropology, social and cultural geography, environmental studies, and literary and art criticism, these books ask what thinking about nonhuman animals can teach us about human cultures, about what it means to be human, and about how that meaning might shift across times and places.

OTHER TITLES IN THE SERIES

Rachel Poliquin, *The Breathless Zoo: Taxidermy and the Cultures of Longing*

Joan B. Landes, Paula Young Lee, and Paul Youngquist, eds., *Gorgeous Beasts: Animal Bodies in Historical Perspective*

Liv Emma Thorsen, Karen A. Rader, and Adam Dodd, eds., *Animals on Display: The Creaturely in Museums, Zoos, and Natural History*

Ann-Janine Morey, *Picturing Dogs, Seeing Ourselves: Vintage American Photographs*

Mary Sanders Pollock, *Storytelling Apes: Primatology Narratives Past and Future*

Ingrid H. Tague, *Animal Companions: Pets and Social Change in Eighteenth-Century Britain*

Dick Blau and Nigel Rothfels, *Elephant House*

Marcus Baynes-Rock, *Among the Bone Eaters: Encounters with Hyenas in Harar*

Monica Mattfeld, *Becoming Centaur: Eighteenth-Century Masculinity and English Horsemanship*

Heather Swan, *Where Honeybees Thrive: Stories from the Field*

Karen Raber and Monica Mattfeld, eds., *Performing Animals: History, Agency, Theater*

J. Keri Cronin, *Art for Animals: Visual Culture and Animal Advocacy, 1870–1914*

Elizabeth Marshall Thomas, *The Hidden Life of Life: A Walk Through the Reaches of Time*

Elizabeth Young, *Pet Projects: Animal Fiction and Taxidermy in the Nineteenth-Century Archive*

Marcus Baynes-Rock, *Crocodile Undone: The Domestication of Australia's Fauna*

Deborah Nadal, *Rabies in the Streets: Interspecies Camaraderie in Urban India*

Mustafa Haikal, translated by Thomas Dunlap, *Master Pongo: A Gorilla Conquers Europe*

Austin McQuinn, *Becoming Audible: Sounding Animality in Performance*

Karalyn Kendall-Morwick, *Canis Modernis: Human/Dog Coevolution in Modernist Literature*

Keith Botelho and Joseph Campana, eds., *Lesser Living Creatures of the Renaissance: Volume 1, Insects*

Keith Botelho and Joseph Campana, eds., *Lesser Living Creatures of the Renaissance: Volume 2, Concepts*

Kaori Nagai, ed., *Maritime Animals: Ships, Species, Stories*

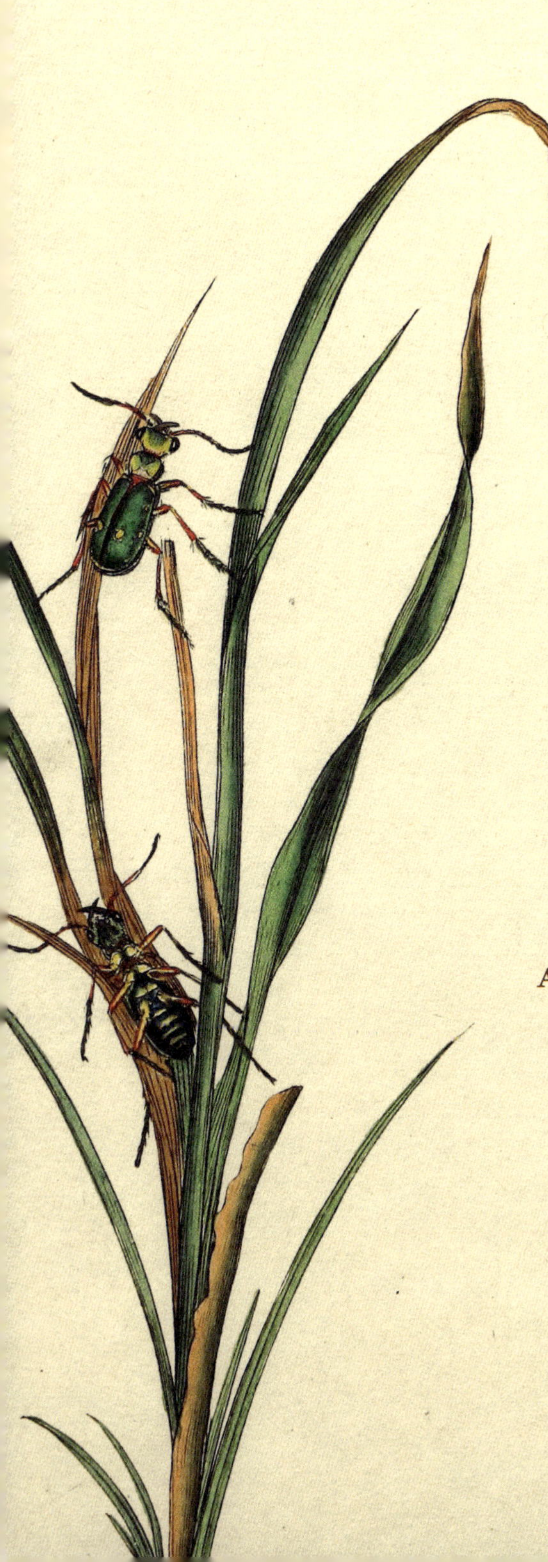

Where the Grass Still Sings

STORIES OF INSECTS AND INTERCONNECTION

Heather Swan

The Pennsylvania State University Press,
University Park, Pennsylvania

Library of Congress Cataloging-in-Publication Data

Names: Swan, Heather, 1968– author.
Title: Where the grass still sings : stories of insects and interconnection / Heather Swan.
Other titles: Animalibus.
Description: University Park, Pennsylvania : The Pennsylvania State University Press, [2024] | Series: Animalibus : of animals and cultures | Includes bibliographical references and index.
Summary: "Celebrates insects, their crucial role in our ecosystems, and people working to preserve biodiversity"—Provided by publisher.
Identifiers: LCCN 2023046845 | ISBN 9780271096957 (paperback)
Subjects: LCSH: Insects—Ecology. | Insects—Conservation. | Biodiversity conservation.
Classification: LCC QL496.4 .S83 2024 | DDC 595.7—dc23/eng/20231026
LC record available at https://lccn.loc.gov/2023046845

Printed in Türkiye
Published by The Pennsylvania State University Press,
University Park, PA 16802-1003

10 9 8 7 6 5 4 3 2 1

The Pennsylvania State University Press is a member of the Association of University Presses.

It is the policy of The Pennsylvania State University Press to use acid-free paper. Publications on uncoated stock satisfy the minimum requirements of American National Standard for Information Sciences—Permanence of Paper for Printed Library Material, ANSI Z39.48-1992.

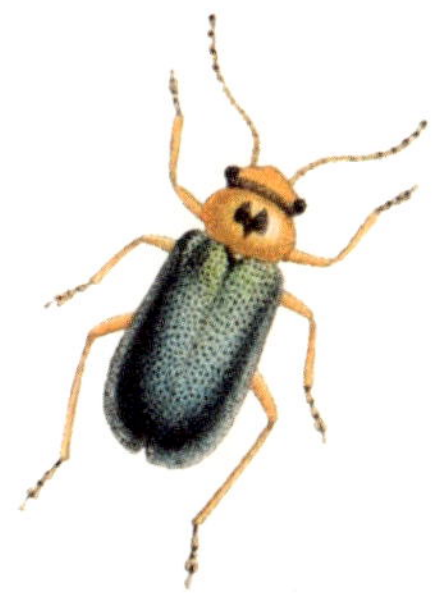

for the myriad
marvelous beings

If all mankind were to disappear, the world would regenerate back to the rich state of equilibrium that existed ten thousand years ago. If insects were to vanish, the environment would collapse into chaos.

—E. O. WILSON

You can't have hope without creativity.

—ARTHUR KDAV

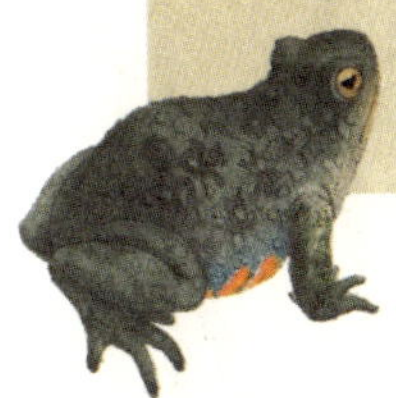

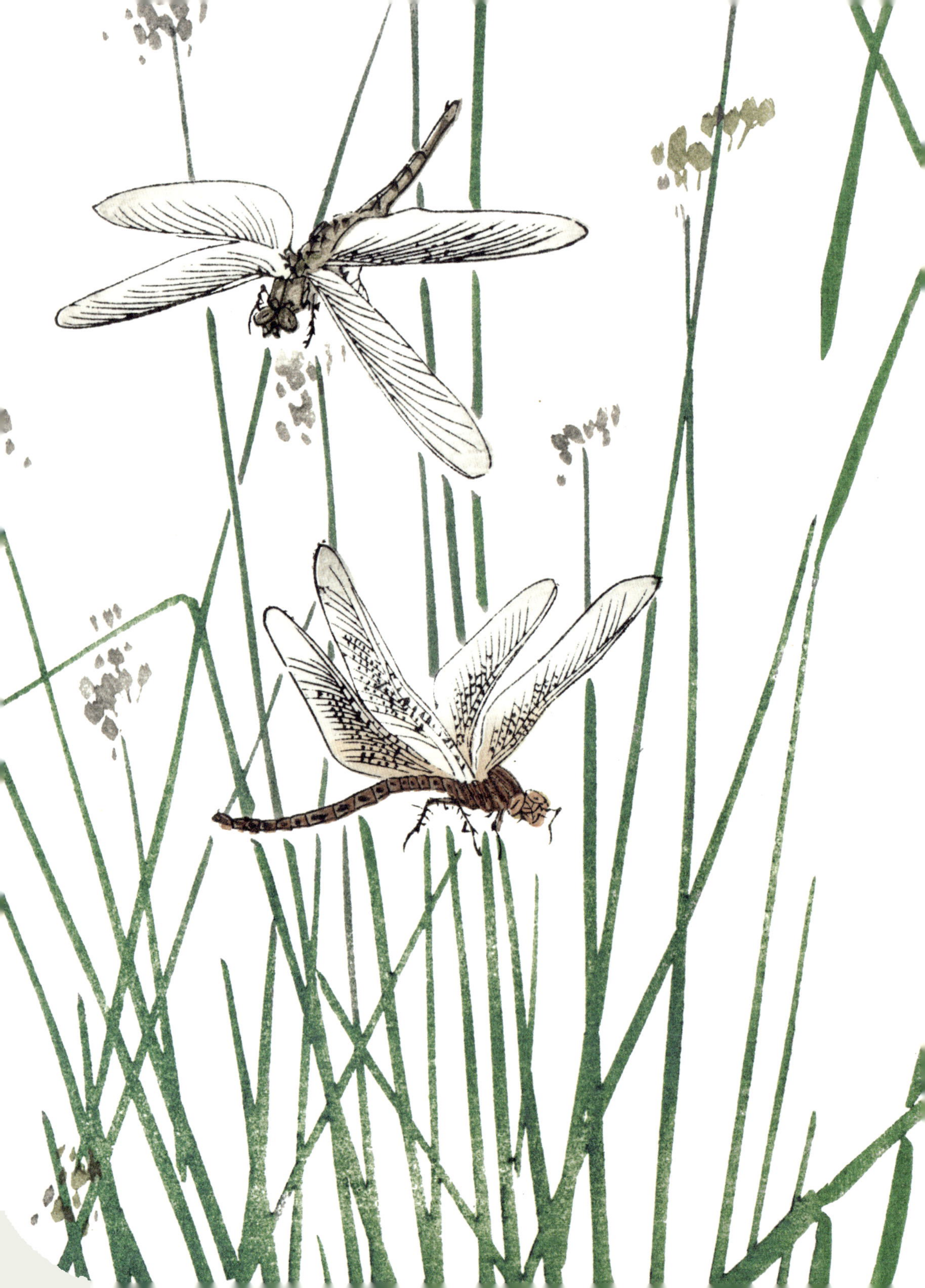

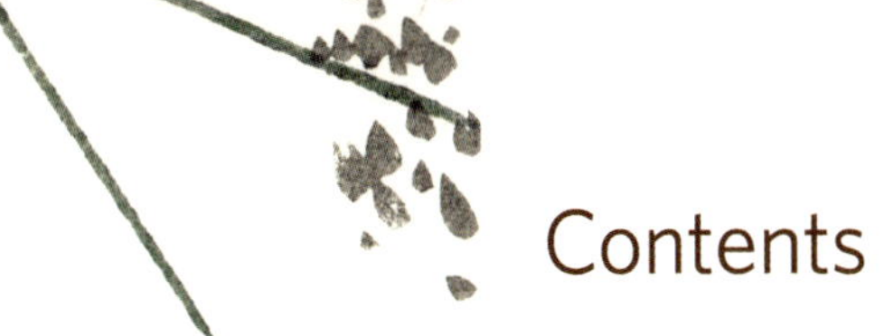

Contents

Acknowledgments

In the same way a singing landscape exists only because of a wide range of plants, insects, birds, amphibians, and mammals, this book would not exist without an amazingly diverse group of passionate people who share a desire to create a more sustainable, just world.

I am grateful to the University of Wisconsin Department of English and the Nelson Institute for Environmental Studies for their ongoing support and to the Global Health Institute for the grant that helped support my research in Colombia. Several chapters of this book (or versions or sections of chapters) first appeared in the following journals: *About Place*, *Belt Magazine*, *Catapult*, *Edge Effects*, *Emergence Magazine*, *Minding Nature*, and *One Art*. I have had the luck to work with some very insightful editors, including Seana Quinn, Tom Jeffreys, Ryan Schnurr, Addie Hopes, Clare Sullivan, Marisa Lanker, Charles Carlin, and Megha Majumdar.

Thank you so much to the inimitable artists: Jenny Angus, Emily Arthur, Lea Bradovich, Susan Carlson, Liz Anna Kozik, Edouard Martinet, Rosalind Monks, Claire Morgan, and Amy Spassov for contributing their wondrous insect artwork to this project.

I am incredibly grateful to the inspiring people I met on this journey who generously spent time with me and patiently taught me about their work. Thank you to Mel and Vickie Gratton; Jason, Jennelle, Myra, Isaac, Genevieve, and Nanka Thimmesch; Josh and Kerstin Mabie; Ivan Lozano; Devin Edmonds; María Luisa Hincapié, Luis Carlos Peña, Mar, and Sol; Ian Boyden; Peter Oboyski; Jeremy Hemberger; Claudio Gratton; Mike Ulyshen; Ernesto and Lourdes Campo; Vicko Castro; Veronica Hinke; Jerry Hoffman; Michelle Pearion; and Corey Searles and the Searles family. Also to Andrew Mahlstedt for hosting me in Colombia and for sharing his magnificent stories, and to Joe Meisel and Catherine Woodward for introducing me to the magic of Ecuador and the great work of Ceiba.

This book came to life in great part because of the example, encouragement, and feedback I had from a stellar community of writers and thinkers, including Daegan Miller, Cherene Sherrard, Amaud Jamaul Johnson, Gavin Van Horn,

Nickole Brown, J. Drew Lanham, Nancy Langston, Catherine Jagoe, Stella Nelson, J. L. Conrad, James Crews, Mary Fiorenza, Peter Boger, Kate Veira, Nathan Jandl, David Zimmerman, Gwenola Caradec, Jesse Lee Kercheval, and Judith Claire Mitchell. Thank you to Laura Reed-Morrisson for her very keen edits and suggestions on early drafts of this book. I am especially indebted the erudite and generous Gregg Mitman, whose creativity, in-depth research, and ethical compass continue to inspire me, and who convinced me to keep writing about insects, and also to my brilliant and visionary editor, Kendra Boileau, who believed in this project from the very beginning.

Huge thank yous also to Jackson, Donovan, Ben, Anna, and Rick Waters, who prodded me endlessly to keep writing this book; to Joe Parisi for making me laugh on a daily basis throughout the pandemic; to Deb Sullivan for falling in love with bugs with me when we were small; to Lynne, Robin, and Leane for the dancing when we needed it, and to VO5 for making the music; to Hope Henley for your gentle wisdom and Mare Chapman for the tools of mindfulness; and to Tom, Susan, Jack, Josephine, Ike, and Fran for the nights of fires and peach pie while fireflies lifted out of the grass. I am honored to be the child of the artists Stephanie O'Shaughnessy and the late Arthur Kdav, both of whose lives were built around art and nature, which shaped how I see the world. Thank you. My life has been utterly changed and made beautiful by the love, the curiosity, and the sharp wit of my children, Maia and Elijah. Thank you. And to giant-hearted Drew Szabo, who has nourished me in so many ways, thank you. And to all the marvelous more-than-human beings around us, thank you. I'm so lucky for this rich life. Thank you, thank you.

Introduction

> We all share the same water, the same air, and the same earth.
>
> —J. DREW LANHAM

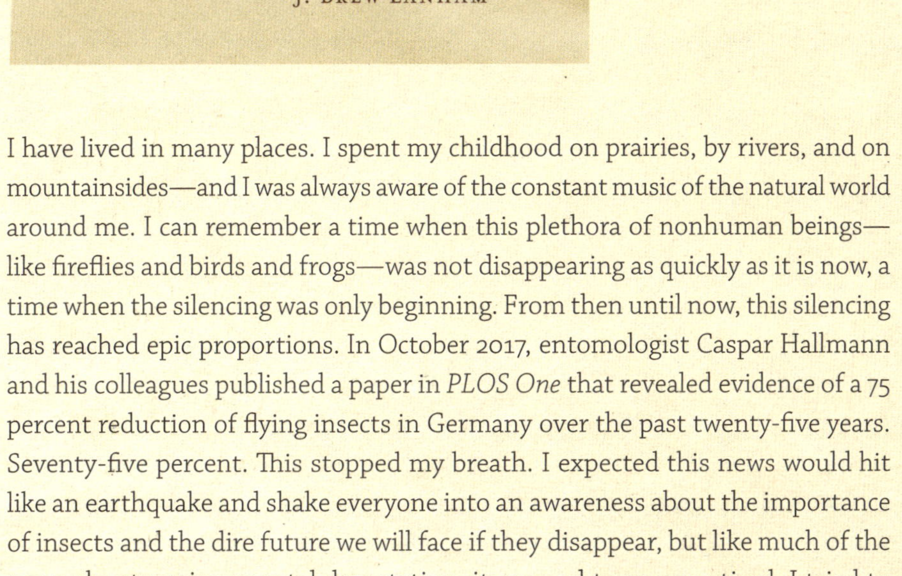

I have lived in many places. I spent my childhood on prairies, by rivers, and on mountainsides—and I was always aware of the constant music of the natural world around me. I can remember a time when this plethora of nonhuman beings—like fireflies and birds and frogs—was not disappearing as quickly as it is now, a time when the silencing was only beginning. From then until now, this silencing has reached epic proportions. In October 2017, entomologist Caspar Hallmann and his colleagues published a paper in *PLOS One* that revealed evidence of a 75 percent reduction of flying insects in Germany over the past twenty-five years. Seventy-five percent. This stopped my breath. I expected this news would hit like an earthquake and shake everyone into an awareness about the importance of insects and the dire future we will face if they disappear, but like much of the news about environmental devastation, it seemed to go unnoticed. I tried to pacify myself with the thought that perhaps the decline in Germany stood alone.

Shortly after reading that paper, as if to quell my optimism, Jacob Mikanowski published a piece in the *Guardian* asserting that the German study was not unique. According to his sources, insect habitats—and insects themselves—were disappearing at alarming rates all over the globe. Many of us are familiar with what experts have called the Sixth Extinction, but one aspect of this dramatic shift that goes largely unnoticed is extinction in the class Insecta. In the piece, Mikanowski evokes the words of renowned entomologist E. O. Wilson, who argued that without insects, humans would only last a few months. Entire ecosystems would quickly collapse. For me, the news of this rapid and impressively large decline of insect populations felt apocalyptic. I wanted to know more about our relationship with insects and what was making this shift happen.

As a child, I thought of all living creatures as part of my extended family. Several years after my father left my mother, and a month after her artist's residency in Colorado had come to an end, my mother took her two daughters—my little sister, Anna, six, and me, eleven—and her dog back to the Midwest, where she had lived when she was still married. Life in cooperative art centers had been rich in many ways but did not produce much money, so when some friends offered to let us live in a barn on one of their unused properties, she took it. At the time, I did not realize that we were living in poverty. Poverty was never something I really thought about as a child. Owing largely to my mother's ingenuity, my life seemed rich in many nonmonetary ways. The reality I lived in was the only one I knew, and so it never struck me as unusual, and certainly not sad, though when I tell stories to my daughter now, she often uses those words. But my mother has a shining, relentlessly optimistic spirit, and to her, living in a barn seemed like a rare opportunity to experience something unique and adventurous—and its being free made it especially attractive. So we moved into the barn.

The barn nestled between two thickly wooded hills, divided at the base by a small creek that ran fast after big storms but was otherwise a trickle. Coyotes, raccoons, owls, rattlesnakes, deer, and even an occasional bobcat resided in those woods. The barn seemed like an anomaly, as it was not surrounded by flat farmland. Years prior, it had been useful to a neighboring farm, accessible through the narrow valley, but it had been long abandoned. My mother's friends had bought it thinking that turning a barn into a living space would be a great architectural project, but they discovered they did not have the time or money to actually fix it up.

The design of the barn was traditional for the Midwest. The lower part of the barn, the basement, was made of stone and concrete, a space for cattle, sheep, or hogs, with large openings for air flow; the bulk of the barn, the upper level, was a huge open space once used for storing hay bales, with plywood floors, knotty-plank uninsulated walls, and thirty-foot ceilings. We lived on that level. The one untraditional feature of the barn was the addition of some windows. On the side that faced the valley, three four-by-eight-foot holes had been cut into the side of the barn and plate glass had been installed. The opposite wall also had a couple of finished windows. But not all of the holes had been filled with glass. One particular hole, a four-by-four-foot square up close to the eaves, was open to the evening air. My mother erected a couple of dividers made out of two-by-fours and drywall to create two ceilingless bedrooms, but the living space was mostly open. A big table near one of the large windows became the hub, the spot where we gathered for art projects and meals. During the day, the tractor-sized sliding door was open. My mother hung up a hammock between two of the giant structural beams, which was a great place for reading. There was no running water, but we did have electricity available, and extension cords allowed us to have lights at night, a boom box for music, a small TV that got two local channels, and a tiny refrigerator on the first level, where the livestock had lived. My mother's pottery wheel and the portapotty also resided on the ground floor. As kids, Anna and I settled into this new spot with ease. We had moved so often that there was little stress about sleeping in a new place or keeping clothes in boxes.

The one feature of the barn that my mother recognized as distinctly different from other places we had lived was the number of nonhuman creatures living with us. Because the barn had so many unclosable openings to the outside world, the communities of creatures living in the surrounding woods frequently came into the building. My mother did not even raise an eyebrow at the mice, but the raccoons coming in for an easy snack unnerved her a bit more. I never saw a snake in there, but I would bet they traversed the basement, as we often saw them outside. My sister and I spent as little time in the basement as possible because it did seem cold and damp and slightly creepy. But the bats and insects upstairs did not bother us much at all, in part because of my mother's strategy of naming everything.

When my mother realized that there was no way to control the number of wasps and June bugs and bats with whom we cohabitated, she decided to prevent her children from fearing them by consciously embracing them as part of the family. I realize now that she must have been influenced by E. B. White's *Charlotte's Web*. I don't remember her ever calling a spider anything other than "one of Charlotte's cousins." Winthrop the Wasp and his friends needed to be given space because of their sensitive temperaments, and we were directed not to swat at them but to move slowly, so as not to startle them; we were never stung. We understood that Junie the Junebug and her friends were not very graceful fliers, and if they ran into you, you shouldn't freak out but just pluck them off your shirt and take them back outside. Filmore the rabbit and his babies would have been most welcome inside, but they never seemed to want to move in. And the bats—and there were a lot of them—were all related to Angel.

Every night at dusk, the bats would fly in through the highest opening in the barn that had no window. My mother, my sister, and I all had very long, curly hair, and although the likelihood of a bat getting stuck in our hair was probably quite low, my mother had us put on hats and tuck our hair up inside them. "Time to put on your hats, girls!" she would say cheerily each night as the tiny black bodies started flying erratically around us. They came in to eat the bugs whirring around our table lamps. I have no memory of being bothered by mosquitoes in there, most likely because Angel and her relatives feasted on them. The electric thrill of having them sweep past your head never ceased, though. Once in a while a very brave friend would stay overnight out there with us, and I remember us being side by side in my makeshift bed with the covers up to our necks waiting for the bats to swoop down near our heads, and when they did, we would squeal and bolt under the covers, giggling in a combination of terror and elation that most people only experience at haunted houses set up for Halloween. But we knew not to be truly scared, because it was just Angel and her relatives, after all.

My mother's creature-naming strategy helped solidify my belief that all of the nonhumans around us have amazing, complicated lives worthy of respect and honor. The foundation of this idea started even earlier for me, when my parents were still together and we were living next to a prairie where I spent days with my dog, inventing stories about all of the insects and toads and birds that lived there. The fact that all of these beings did not share a language with me did not strike me as something indicative of their inferiority. If anything, I thought humans were the ones with limitations. To this day, it feels like a gift to be in the presence of wild things. One of the most powerful aspects of this kind of experience for me was the sound.

I remember so clearly the sounds of frogs singing on spring evenings and how that ringing in the air helped me sleep, how the crickets' thrumming soothed me, how the cicadas' relentless vibrations seemed to clearly articulate the intensity of the heat and humidity of summer afternoons. Wrens, robins, mourning doves, and cardinals created an inviting chorus early each morning. And those sounds allowed me to define spaces: singing meadows, creeks, hollows. Years later, when I began keeping honeybees, the sounds of the hive gave me an immense sense of peace—as they do to this day. A landscape in this part of the United States being devoid of sound is terrifying to me. And more and more, as Rachel Carson so clearly imagined and illustrated in her opus about DDT, silent springs are now becoming a reality. Wild things are disappearing all over the world.

Part of the lesson of living in that barn was that not all of the creatures with whom we coexist are ones that make us feel comfortable. The wasps and the bats were part of the fabric of life, and we were taught to respect them, not to fear them or try to destroy them. Once during one of my classes in college on a hot, humid afternoon, a very practical application of this lesson came in handy. The professor had propped a door open to let a breeze in. Suddenly a young woman started screaming, jolted out of her seat, and started pointing breathlessly to a wasp that had landed on the floor near her desk. Others responded similarly, jumping up from their desks and backing away. The class was devolving into mayhem. I suppose it may seem self-aggrandizing to tell this story now, but at the moment, it was simply an obvious course of action. I knew that the wasp would become alarmed and eventually sting if we all became frantic, so I walked over to it and crouched down, lay my hand down flat in front of it until it took a few cautious steps onto my palm, then carried it out to freedom. The professor expressed gratitude for making it possible for the class to resume. That wasp had no interest in hurting me. It was just in the wrong place at the wrong time. So often the knee-jerk response to finding something that is not convenient or pleasant for us humans is to kill it. But our blithely destructive practices have terrible consequences. Pesticides and herbicides may be eliminating the things we don't like, but they are also destroying ecosystems.

In 1902, a Russian biologist named Pyotr Kropotkin became frustrated with the co-opting and oversimplification of Charles Darwin's idea of evolution, which had been distilled down to the phrase "survival of the fittest." In direct response to a writer named T. H. Huxley, who was using this phrase to justify all kinds of societal oppression and atrocity against the "weak," Kropotkin built on a lesser-known imperative of evolution, one acknowledged by Darwin: the existence of mutual aid. In the first chapter of Kropotkin's book entitled *Mutual Aid*, which expounds upon his understanding of this phenomenon among animals, he meditates on the cooperation of honeybees. His argument was that the habits and success of the honeybee community refuted the idea that fierce competition, or "struggle," between individuals was the key to survival. He found the opposite to be true and found the idea of violence for advancement repugnant. According to his theory, the firm distinctions between individuals and communities needed to be troubled as we thought about survival as a species. Unfortunately, Kropotkin's critique of Huxley did not get much traction, and the "survival of the fittest" version of Darwin's narrative continues, in most cases, to be the predominant one. This idea fits nicely into a capitalist system but has devastating ramifications as we face the environmental and cultural emergencies of the Anthropocene.

More than one hundred years after Kropotkin wrote, we are seeing the consequences of an unhealthy idea of species fitness and domination. Humans are bearing witness to the rapid decline of bees, all other flying insects, and, arguably, all other nonhuman species. A beekeeper trained in Zen philosophy named Michael Thiele writes about what we must now learn from the honeybees. He pushes beyond Kropotkin's idea of mutual aid within a species and suggests that we acknowledge what the bees recognize and live daily, which is what he calls "interbeing," a term borrowed from the Vietnamese Buddhist monk Thich Nhat Hanh. His notion draws on the idea that pollinators are exemplary in their relationships of mutual aid, incorporating not only other bees but plants, water, and even humans as well. The interbeing concept suggests that the success of one individual or one species requires a transgression of distinctions between the life of one organism and another. This is, of course, ecosystem thinking. Bees could not survive without plants; humans cannot survive without insects, which pollinate our food, or plants, which create air we can breathe; and so on. Many of our current practices in agriculture and resource extraction are contributing to the destruction of this web of relationships.

How can we shift this narrative? The answer, for me, was to learn more about the web of relationships we depend upon and then to write this book, which aims to bring to light the stories of people who have deeply inspired me, who are living this idea of interbeing, who are supporting ecosystems, and who have become insect advocates in our precarious moment. This book is a collection of wonders and sorrows as well as a playbook for the survival of our planet. It integrates scientific data, poetry, art, and the daily practices of farmers, entomologists, conservationists, and artists whose work—seen together in a mosaic—can help us begin to think differently about cooperation across the nonhuman/human boundaries and about how to preserve a healthy, biodiverse planet.

Beginning with a visit to an insect museum curator and then on to meetings with people in Ecuador and Colombia, two of the most biologically diverse countries in the world, and eventually to the midwestern United States, where you can witness the devastating effects of industrial monocropping with agrochemicals, the book illuminates our precarious and crucial interdependence with insects. Each chapter introduces another story about innovative people and projects aiming to preserve an appropriate balance so that insects and humans (and other nonhumans) can thrive.

After each chapter, I include a gallery of artwork. The artists are doing different but equally important work in illuminating our present or our past and often offer pathways to a different future. Some of the work responds to scientific knowledge, such as the prints of the Cherokee artist Emily Arthur, who says of her work, "Art is not in the service of science. Art and science share the responsibility of observation and witness. It is through observation that science gives us proof of our material make up. It is through observation that art gives us material proof of our spiritual make up. Encountering a great work of art or a great leap in science changes our perception; it asks us to see and then to see once again, more deeply." Some of the work reflects forgotten wisdom that needs remembering. Some of the work celebrates the miraculous and the strange. Some offers a vision for the future.

I am indebted to the writings of Rachel Carson, Annie Dillard, Jane Hirshfield, Robin Wall Kimmerer, Elizabeth Kolbert, J. Drew Lanham, Rebecca Solnit, Terry Tempest Williams, and E. O. Wilson, from whom I have been inspired to blend scientific information and lyrical language in a way that invites readers to rekindle their sense of wonder and perhaps experience a sense of urgency to care more deeply about our planet, to work on building kinship with each other and our fellow earthlings. I hope to honor the unexpected and often invisible intimacies of humans and nonhumans in order to help us navigate a better coexistence. I want our descendants to know the magic of standing in a field late at night as the air vibrates with song and fireflies create a tapestry of light.

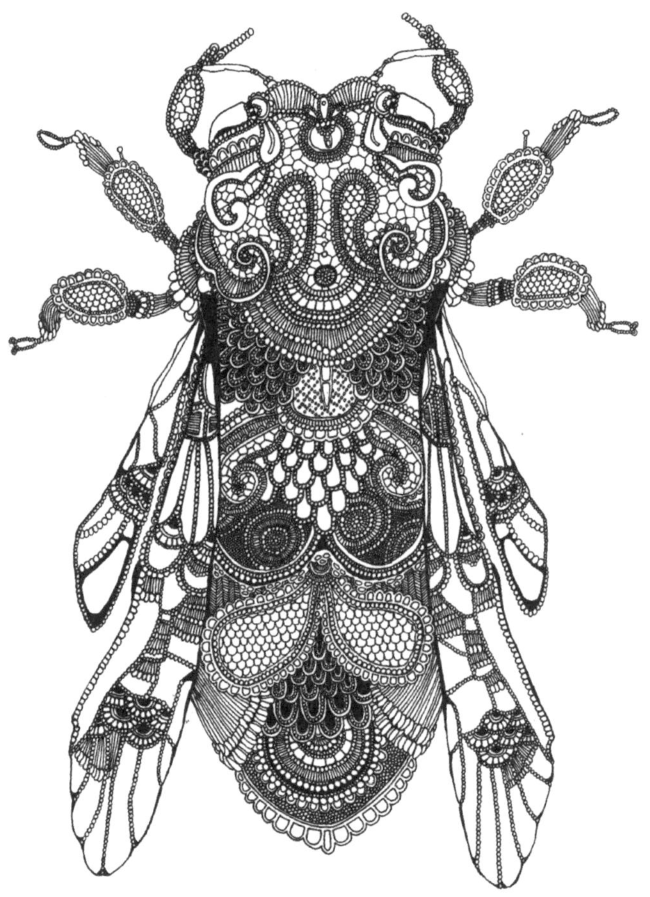

Rosalind Monks, *Honeybee*, 2011. Pen and ink, 297 × 420 mm.

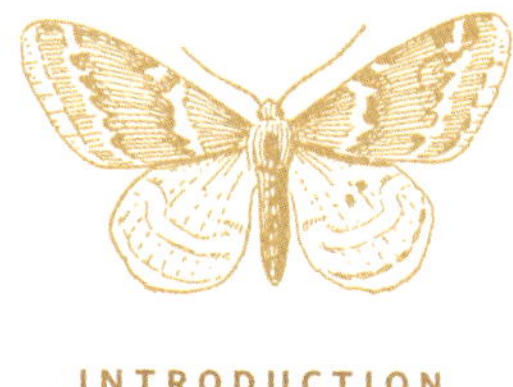

Rosalind Monks and Invisible Intricacies

I believe a leaf of grass is no less than the journey-work of the stars.
—WALT WHITMAN

Every semester at the university where I teach, I open my classes by reading a poem called "The Summer Day" by Mary Oliver, a maven of observation. The poem ends with the famous lines, "Tell me, what is it you plan to do / with your one wild and precious life?" The poem is great for anyone at any age because this essential question reminds us of the precious, precarious nature of our existence and asks us how we want to use the limited time we have on this planet. I repeat this line to my students for emphasis. But I also point to another part of the poem, in which Oliver is describing a grasshopper:

This grasshopper, I mean—
. . . .
the one who is eating sugar out of my hand,
who is moving her jaws back and forth instead
of up and down—
who is gazing around with her enormous and
complicated eyes.
Now she lifts her pale forearms and thoroughly
washes her face.
Now she snaps her wings open, and floats
away.
I don't know exactly what a prayer is.
I do know how to pay attention

What strikes me is that the poet has actually answered her own question with these lines. Her answer for how best to spend our time on this planet is to pay attention. Paying attention to the most fleeting moments or the smallest, most intricate beings is part of what makes her life meaningful. The individual grasshopper she describes is a brilliant example of the immense mystery that the natural world provides to us every day but that we so often ignore or pass by. Oliver's poem makes us stop and notice with her.

When I first saw the artwork of Rosalind Monks, I recognized a kindredness between the poet and the artist. Monks's work also grabs the attention of the audience and does not let us pass by the smallest beings. Her intricate drawings of insects invite the viewer to contemplate the complexity of each unique creature.

Monks grew up in the Alps, surrounded by the majesty of the natural world. Lakes, mountains, and ravines offered the experience of immensity and perhaps something many of us would call the romantic sublime, the experience that fills us with awe and a touch of fear that can result in humility, the recognition that the human being is a small part of a much larger tapestry of life.

But the many creatures who populated those spaces were not lost on Monks. As she hiked and skied through the Alps and traveled throughout New Zealand, she took note of the world around her and wanted to re-create it in her artwork, but in a way that was uniquely hers. Her delicate line drawings depict real-world animals that are highly stylized and often have a mythical presence. Her major in illustration at the University of Brighton created a foundation for her as an artist, and her style has evolved since then.

Monks's *Cicada* especially captivated me. Each line, each tiny detail, every rich imagined space spoke to me about the impossibility of ever completely understanding a being that is so radically different from us. And cicadas themselves capture my attention. My first awareness of cicadas came when I was a little girl. I remember the humid air of July pulsing with an incessant thrum, a sound that was impossible to ignore. The sound of high summer. It was astonishing to me that insects could create such a powerful noise. Tiny organs called tymbals give cicadas this ability. Unlike a cricket, who rubs her wings together, a cicada's tymbals are membranous areas on the abdomen that vibrate from muscle contractions. The shape of the cicada's body amplifies the sound to up to 106 decibels, making it one of the loudest insect sounds. While the cicada noise rang through the air, I would collect the discarded cicada exoskeletons that fell to the ground as the new cicadas emerged and climbed into the trees. The "shells," as we called them, were thin, hollow, cicada-shaped cases left behind; they had every detail of the cicada and were the color of maple syrup.

Annual cicadas actually live underground for approximately five years and come to the earth's surface in batches, so there are cicadas in many parts of the world every summer. But some cicadas have a life cycle of seventeen years, and when they emerge, they do it en masse. This event is hard to miss and has been documented for hundreds of years, a powerful example of the natural cycles we can count on. And yet an article in *Scientific American* recently noted that sometimes the cicadas are emerging early. A cause is not understood. Could our behaviors be affecting them? Is climate change or the chemical changes we are making on the earth affecting them? And how are they experiencing these things?

Shortly after the devastating Fukushima tsunami, earthquake, and subsequent nuclear disaster in Japan in 2011, a short animation, *663114*, was created by Isamu Hirabayashi. The film follows the life cycle of a cicada born into the year of the disaster. The explosions of the nuclear plant are suggested subtly in the film, and when the next generation of cicadas begins its journey,

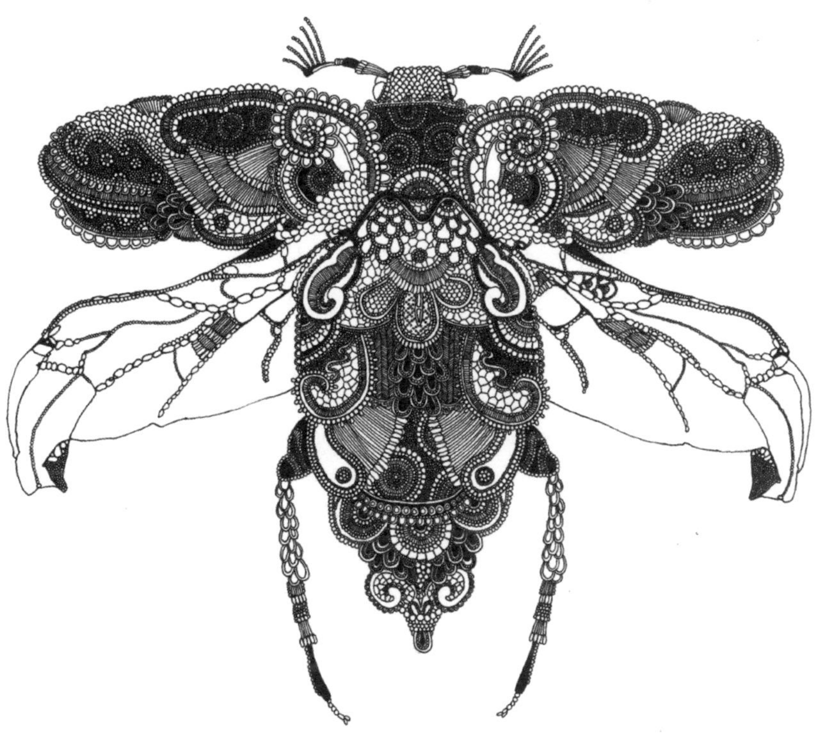

Rosalind Monks, *June Bug*, 2011. Pen and ink, 297 × 420 mm.

the film shows a cicada, utterly changed, disfigured and singing with a garbled voice. We are left to think about what the radiation will do to those systems we count on and what our impact on the earth has been. There is so much about insects we don't yet understand and so much about our manipulation of the planet we may not foresee. And often these insects, and many others, are maligned and exempted from our circle of care.

In her poem "Magicicada Septendecim," Nickole Brown laments that the first sound of her childhood was tainted by the messages of her community, which taught her to shriek "locust, taught to stay inside, to fear / a being that no one bothered to know / had no mouth to bite nor stinger to sting." She sees how it was human manipulation of their habitat that led to the blanket of dead insects that so offended people who lived there.

MAGICICADA SEPTENDECIM

My loves, my loves, what were you ever to me
but my first sound? Not the first sound like the thrum-dum
that first made my first blood and beat overhead
until I was born, but a deeper first thirst, ear-remembered

without memory, down deep enough to configure those tiny
inner bones that let me hear you, a scapulimancy
pumping blood-strong.
And yes, I should have known when you first
wheeled the suburban yard I never noticed as a kid
into something I could no longer ignore,
when you took of the tree's roots and the sun that made
the tree to make that one tree

pure song. I should have known then
how sacred the muscled instrument of your bodies,
but I was taught to call you locust, taught to stay inside, to fear
a being that no one bothered to know
had no mouth to bite nor stinger to sting.

So forgive me if in ignorance I said outbreak *or* invasion,
if I left you to whiten on the hot sidewalk, did not lift
your electric bodies and move them safely aside.
And forgive me when I failed to see
your seventeen years of refuge—not mine
I know but I should have known

how you dug your way out,
what it was to convulse yourself into being
after those years, to leave that mud-labor
lonely behind, to push and push and push
until you split yourself

down the center spine-line, clinging to the shell
of who you once were while you waited for the pale
soft of you to hardshine into a body dark as coal.

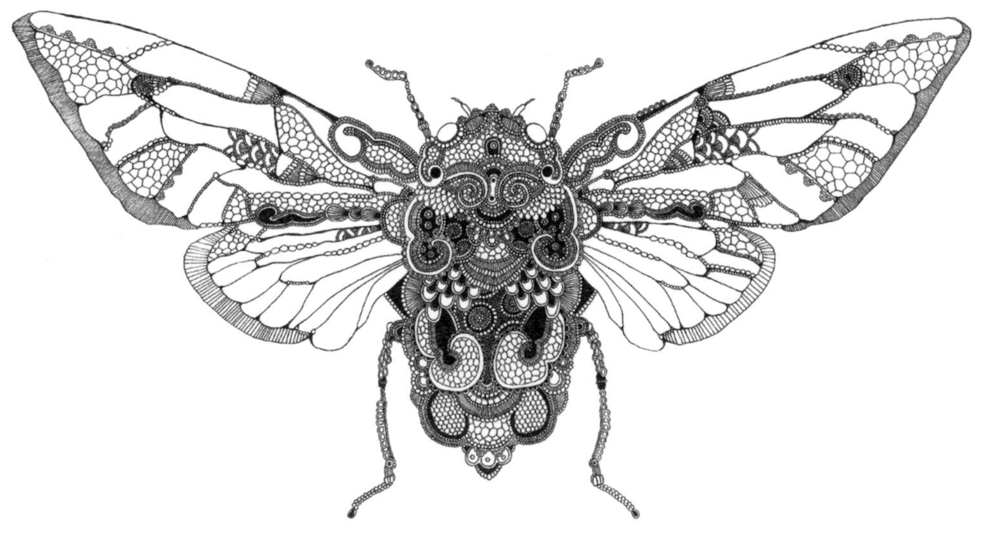

Rosalind Monks, *Cicada*, 2011. Pen and ink, 297 × 210 mm.

You were never metaphor,
my loves, my holy ones, but did you have
something
to teach me, did I have something to learn.

Brown's poem shows us how our childhood sense of wonder and attention can be trained away and shaped by a society that does not recognize the magnificent strangeness of these entities worthy of life. Instead, we have often been trained to fear and to eliminate the things that are inconvenient to us. We can change this. We can realign our thinking.

Rosalind Monks's fanciful designs drawn on these larger-than-life insects are a visual metaphor for all we don't see or observe, a depiction of the wonder we might be missing in the insect world, an invitation to the mystery of how other beings could be experiencing and understanding this planet in a way we never can. Cicadas are just one example. Imagine this: there are bees in Turkey who build their fragile nests out of flower petals, creating the most delicate and beautiful pastel cups for their babies. Honeybees dance maps in the dark for the other bees in their communities to find the fragrant flowers that need pollinating. Fireflies light up! Monks's artwork invites us to look more closely at the world, to pay attention, to investigate even the smallest creatures we encounter during our one wild and precious life.

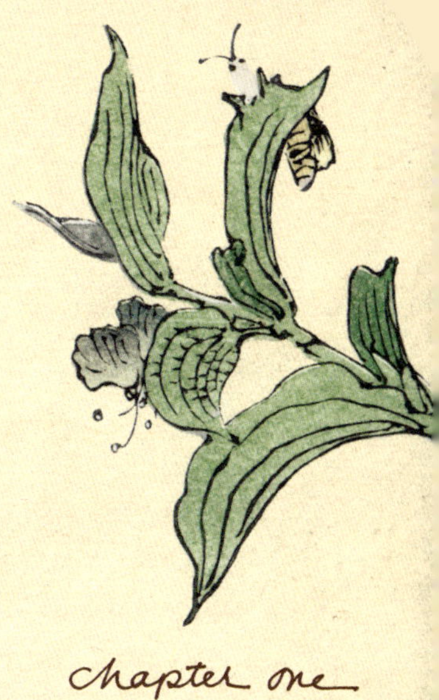

chapter one

Invisible Extinctions

THINKING ABOUT INSECTS AT THE ESSIG COLLECTION

Dyed he is with the
colour of autumn days,
O red dragonfly.

—BAKUSUI

Insects are disappearing at an alarming rate. But how do we know this? One way we know is from the work of people across the world carefully documenting the insects' lives and histories. During a trip to California a few years back, I sought out Dr. Peter Oboyski, entomologist and curator of the Essig Museum of Entomology at Berkeley, who generously offered to talk with me about the role of insects in his life and how he works to understand shifting insect populations over time.

Photos of insects don't generally inspire heart emojis or "OMG. So cute!" comments like those of cats, sloths, pandas, or polar bears. More often, close-up images of insects inspire fear or repulsion, or at best, perhaps, bemusement. So the decision to spend your entire life looking closely at bugs, preserving their delicate bodies, and trying to inspire others to care about them may seem a bit unusual. But Peter Oboyski has made just this choice. The entomologist fell in love with insects long before he became the curator of the Essig Museum, a collection of over five million physical specimens—including approximately thirty-five thousand species of insects and spiders—that he is in the process of turning into a digital collection. The database will amass images of even more insects, as he is coordinating with seven other collections across the state of California. Oboyski has dreams of getting the public more excited and connected to this collection, and to insects in their everyday lives, by making it more accessible.

The Essig Museum of Entomology is housed in a modest room in the lower corridors of the monumental Valley Life Sciences Building at Berkeley. While the *Tyrannosaurus rex* skeleton at the entrance is certainly a head-turner, I would argue that the insects found in the drawers of Essig are absolutely as stunning. When I visited, Oboyski first opened a drawer to reveal spindly brown stick insects as long as my hand and a leaf insect, *Pulchriphyllium giganteum*, from the same order (Phasmatodea) that was just as long and three fingers wide. It looked to me like a nature project made by a kindergartener—like a bunch of leaves glued together in a symmetrical pattern. What an ingenious act of camouflage! Another drawer held rows of the radiant blue morpho butterflies so often featured in Costa Rican tourist brochures, the color of their wings reminiscent of the cerulean blue of the coastal waters of the Caribbean, an effect made possible by light reflecting on the layers of thousands of tiny scales that make up a butterfly wing. Another drawer held rows and rows of jewellike beetles, some of them blue-black with enormous horns, others like dime-sized iridescent emeralds. It is easy to understand why the Egyptians immortalized them in the form of scarabs. In yet another drawer was another clever disguise: the elegant, velvety-brown owl butterfly's ornately patterned wings feature two large black spots that could easily look like an owl's eyes to a confused predator. Then drawer upon drawer of bees, grasshoppers, damselflies, and dragonflies of many colors and sizes—and thousands more. Oboyski has personally collected an impressive number of Indonesian Lepidoptera (moths and butterflies). Each insect is carefully mounted on a pin, which punctures its body slightly to the right of its center line so that one side of its symmetrical body remains fully intact. Beneath each body is a tiny card that most often states its formal name as well as where and when it was found. The collection is a true treasure trove for insect researchers.

The morning I traveled to the museum, a 4.5-magnitude earthquake shook the Berkeley hills. I immediately thought of the insects in the glass-covered drawers. I was not the only one. When I arrived at the collection, Oboyski and a graduate student stood by the shelves—which contained thousands of insects as well as labor-intensive entomological data collected over scores of years—discussing earthquake protection. Oboyski was planning to make improvements to make sure that the drawers could not be shaken easily out of their cases. Unless "the big one" hits, he said. There was no way to really protect the insects from that. But the vulnerability of these physical specimens is only part of why he is interested in digitizing his collection.

Several years ago, Essig became the lead institution on a project called CalBug, an NSF-supported collaboration by eight major terrestrial arthropod collections in California to make one large digital collection—a daunting, almost unfathomable, task. Only 10 percent has been digitized at this point, and the CalBug collection already includes over a million specimens. Oboyski currently has many volunteers working on this project, including undergraduate students and citizen scientists. Aside from the fact that California collections are threatened by earthquakes, why would digitizing this collection be so important? Part of Oboyski's interest comes from his desire to understand the relationship of insects to their environment over time, an issue that seems particularly important as the earth is experiencing such rapid and immense transformations, including mind-boggling numbers of extinctions and habitat loss due to land-use change and climate shifts. Fluctuations are always happening in ecosystems, but what is natural or normal at this point? And should we be concerned?

Oboyski's own path began with his childhood fascination with the natural world. Busy bird feeders peppered his Connecticut yard, allowing him to observe an array of northeastern songbirds. He knew the names of trees. He intently observed tadpoles metamorphosing into frogs. The natural world captivated him. When he went to college, he intended to pursue electrical engineering, but he soon began taking biology classes "on the side" and realized that, in fact, biological sciences were his passion, so he switched majors. In his last year, he took a class in entomology and fell in love. It was like discovering a whole new world, he said. All around him, he suddenly saw things he had never seen before, things that most people never saw at all.

Oboyski moved to Oregon to do his master's degree and began thinking about habitat and insect life by engaging in a river habitat continuum project, looking at changes in insect populations in riparian zones in the Cascades. He was particularly interested in the differences in wetland insect activity in rivers and streams, which varied in large part because of the amount of solar input and algae growth. After completing the MS degree, he worked on a moth survey comparing clear-cut and old-growth forest regions. The results showed that while there was high species diversity in clear-cut areas, clear cutting forests created homogeneous insect communities across the forested landscape, which at a broader scale reduces overall diversity—important information for those working in forest ecology or on conservation projects. Eventually he earned his PhD at Berkeley, where he focused more deeply on moths, particularly on tiny moths who developed between the layers of a leaf. His project in Hawaii, before he began the PhD, however, zeroed in on one issue that I was hoping to discuss with him: insects as crucial parts of ecosystems.

Between 1996 and 2000, Oboyski participated as an entomologist in a study examining the causes of the steady disappearance of the critically endangered palila, a golden-headed honeycreeper native to Hawaii. Why would a person who studies insects be called in on a bird project? Because palilas depend on moths for food. Palila parents feed their chicks the caterpillar of a specific moth in the *Cydia* genus. This particular moth lays its eggs in *māmane* trees, and when the caterpillars emerge, the palilas harvest them. The research team discovered that a non-native wasp, intentionally introduced as an agent for biological control, was also preying on the *Cydia* moths. Additionally, habitat reduction was creating a crisis for the moths. As the *māmane* disappear, the moths disappear. As the moths disappear, the birds disappear. Here was a clear example of how a variety of species can be so interconnected. Ecosystems are so fragile, I remarked.

Oboyski was quick to point out to me that "fragile" might not be the best word. In fact, ecosystems constantly fluctuate and are quite dynamic. They can recover from subtle changes, but collapse does happen when there are too many. Oboyski used one of Paul Erlich's analogies, which likened a working, healthy ecosystem to an airplane. In Erlich's analogy, the species in an ecosystem represent rivets on the plane. If you know that one rivet is missing from an airplane wing, you'd probably still get on the flight, but if you know that a lot of rivets are missing, you know that the plane is much more likely to fail. Oboyski noted that landscape changes destroy more and more insect habitats, and urbanization eats up many landscapes. These changes can certainly be devastating to ecosystems. I thought about bulldozers and chainsaws, the felling of trees, the endlessly expanding pavement, mountaintop removal mining, the destruction of marshes and fens, housing developments, the warming world, and rampant pesticide use. Where were the creatures to go? And which ones were left? I asked Oboyski his thoughts on how best to investigate insect populations through time.

The Essig Museum and other collections like it are a good start, but the data can be overwhelming, as they are not always gathered in geographical clusters. Sometimes specimen cards are hard to read. Sometimes the location might be vague. Sometimes, however, you get a gift, a key to a treasure chest. Oboyski told me that Dr. Joanie Ball-Damerow found just such a key when she discovered the work of a naturalist named C. H. Kennedy, who sampled Odonata (dragonflies and damselflies) in specific regions of California in 1914 and 1915. Ball-Damerow, then a PhD student in entomology at Berkeley, found Kennedy's work and revisited forty-five specific sites where he had sampled for dragonflies. At those sites, a whole century later, she surveyed the current species richness and habitat. She and her team examined "changes in species occurrence rates, taxonomic richness, and biological trait composition in relation to climate changes and human population increases." The results described how the dragonfly populations had altered over time as habitat changed or vanished. Some species, the "generalists," were more successful at survival than the "specialists" had been. As the name implies, specialists require a unique set of food sources and environmental parameters. Urbanization and water use/shortages take their toll, but there is also evidence of resilience and flexibility.

A couple of years ago, my son, Eli, assisted on a dragonfly survey coordinated by Bradley Herrick from the University of Wisconsin–Madison Arboretum. The project aimed to gather data about dragonfly populations and habitats from approximately fifty ponds in urban and rural wetlands in southern Wisconsin. Included in the project were drainage ponds near industrial areas, ponds in the UW Arboretum that were surrounded by restored prairie, and large natural spring–fed ponds. I accompanied him on a few occasions, climbing through brush higher than my head or leaping over rock barriers and fences to reach our destinations. As Eli sampled water turbidity and took note of wind speed and temperature, I observed the phenomenon of dragonfly flight. In a feat of absolutely brilliant engineering, a dragonfly can hover, fly backward, make incredibly swift changes in direction, and reach speeds of thirty-five miles per hour as it beats its four wings approximately thirty times a second. These adult insects are predators and feed on mosquitoes and other smaller insects, and their larvae eat mosquito larvae as well. One dragonfly can eat anywhere from thirty to one hundred mosquitoes a day. Dragonflies also provide food for fish and frogs, among others. Eli taught me that the large blue-green dragonflies were called green darners, the dark brown–bodied ones with white markings on their wings were aptly named white-spotted, and we even saw a smaller, flashy red dragonfly called a scarlet skimmer. I admit I was shocked that all of these varied wetlands seemed to have quite healthy dragonfly populations. Of course, I had no idea which species were missing, but I saw how urban wetlands could potentially provide habitat for species that were experiencing reduction of their natural habitats. There are many variables—including, crucially, water quality. Herrick's study is yet to be unveiled but will undoubtedly help our communities think about how to take care of those spaces in order to provide a place for the generalist species, at least, which can be more flexible about their living arrangements. In the few days that I accompanied Eli, I gained a huge respect for the work of entomologists, who must have incredible powers of patience and observation.

This keen power of observation is something I also found in the poetry of Catherine Jagoe. Her poem "Cherry-Faced Meadowhawk" is not only a vivid description of the physical qualities of a dragonfly. It is also a love poem to a beautiful bright red being whose presence is helping her carry on.

CHERRY-FACED MEADOWHAWK

Sympetrum internum

Small carnivore soaking up late sun,
is that a ruby or a bloodspot on your wings?
As our northworld tilts toward the dark,
you weave air and gold October light
with your scarlet spindle, your slim
wand that's spun and spun inimitable spells
through eons of lost time, primeval
aviator, here before the birds, before
the dinosaurs. Your stigmata are not
wounds, but mystic marks of grace.
You snatch your few short weeks
of bliss, fighting and fucking on the wing,
trapeze artist decked in cherry goggles
while the leaves loose and release
themselves, locusts' ocher minnows,
ashes' wine-dark ovals, October
with all its losses, trees' great galleons
ablaze, while all along the globe
spins and spins around the sun, regardless
of the gouts of blood, canary, flame,

the flocks of leaves scudding and drifting
like snow that's all too soon to come.
Meantime, you bask in the slant sun's
sweetness, last life-giving warmth
as I feast on your fire-engine red, infuse
it into my bones to tide me through.

Before I left the Essig Museum, I asked Oboyski about his dreams for his future work. In addition to digitizing his collection for researchers, he is very interested in outreach to larger communities. He and his team already have events like their Butterfly Workshop, when the collection is opened up to the public and experts give talks about life cycles, host plants, behaviors, and identification. He wants everyone to discover the amazing world of insects. I thought of my own journey with insects and how the honeybee had led to my larger interest in these creatures. I have certainly come to understand their enormous and crucial role in our ecosystems, and this has galvanized my desire to work for their survival. I imagined the same might be true of visitors to Oboyski's workshop.

Oboyski's other big dream is to create an app that will help people identify insects—or simply *see* them—as they are out hiking or walking down the street. He thought this might create a new kind of mindfulness about insects around us, for both young and old. Perhaps with this increased awareness, we could more easily shift our consciousness about human impact on the earth and then move forward a bit more gently and be more cautious about destroying habitat after habitat. I asked him what advice he had for a young entomologist. "Watch," he said. "Watch the way they behave. Watch how a spider weaves a web. Watch where that beetle is going. Watch the way species interact. You don't really need to catch them. You can learn so much by just observing." What we can discover is nothing short of spellbinding. After my meeting with Peter Oboyski, I left the humble room and stepped out into the bright afternoon, determined to keep my eyes open.

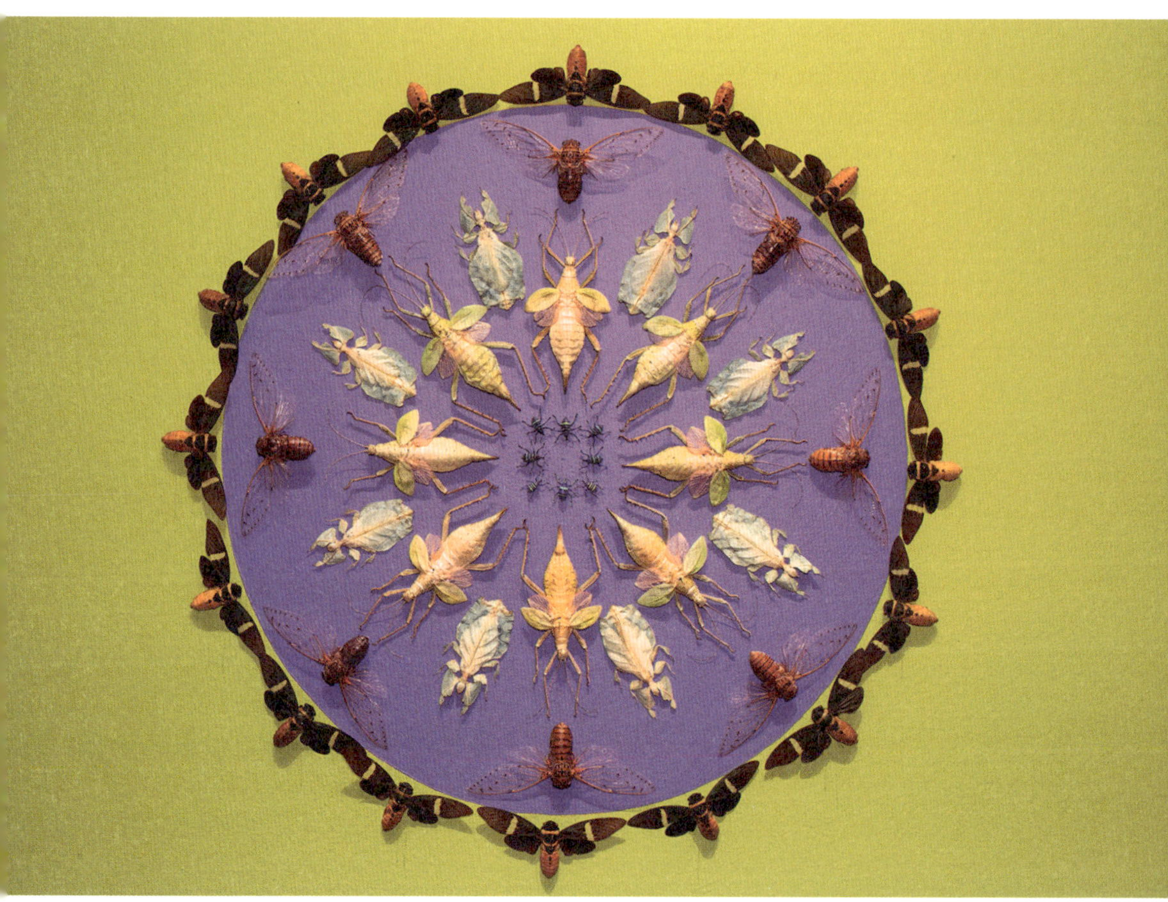

Jennifer Angus, *The Grasshopper and the Ant*. Installation view: insects, taxidermy, view 1. Museum of Fine Arts, St. Petersburg, Florida, 2019.

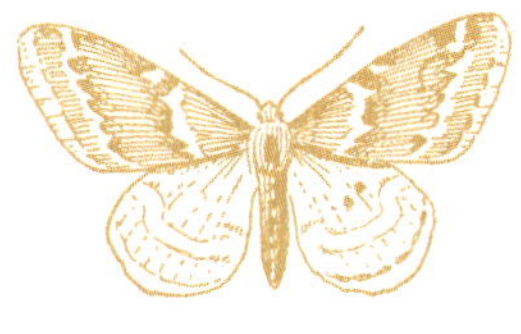

Jennifer Angus

Patterns

If you notice anything,
it leads you to notice
more
and more.
—MARY OLIVER

What if you walked into a room in an art exhibit and realized that the walls covered in vibrant, intricately patterned wallpaper were actually covered in bugs? Would you run out of the room because the mere idea seems like something out of a horror movie or a science fiction novel? If you realized that the insects were all dead, would you want to stay and, out of curiosity, look a bit more closely? Would you feel some amount of sadness? If you stayed, maybe you would even recognize that some of the patterns were made with cicadas, some with pink-winged grasshoppers, some with weevils, and others with leaf insects. Perhaps this would fill you with disgust—or perhaps you might even find the insects up close, dare I say it, beautiful?

Jennifer Angus covers the walls of her exhibits with patterns made with insects because she knows the range of her audience's reactions. The designer and installation artist has always loved patterns, especially on fabric. She first noticed the incorporation of insects as a design element in tribal clothing when she was at an artist's residency in northern Thailand. "I'm a magpie," she says, "attracted to shiny things." A garment studded with the glimmer of metallic beetle wings caught Angus's eye. With their bright colors and strange shapes, insects could be quite inspirational to a designer.

Later, while Angus worked in a studio in Japan on the outskirts of Tokyo, near Mt. Fuji, two young boys, Yoshi and Nori, came to visit her every day and would bring their pet beetles. She remembers that they began to bring her live insects when she showed interest. Years later, she would begin using her now-vast insect collection for installations so that she could impress visitors with the creatures' complexity and plenitude. Angus wanted viewers "to think about them . . . and to say 'wow.'"

In 1665, a scientist named Robert Hooke created the first detailed drawing of a flea in a book he called *Micrographia: or*

Some Physiological Descriptions of Minute Bodies Made by Magnifying Glasses with Observations and Inquiries Thereupon. Hooke's were the first drawings of magnified objects to be presented to the public. Critic Tom Lubbock writes that Hooke "was curator of experiments of the Royal Society, and the range of his work is remarkable: microscopy, elasticity, gravitation." In the opening pages of his book, Hooke writes a slightly tongue-in-cheek message to the king: "I here presume to bring in that which is more *proportionable* to the *smalness* of my Abilities, and to offer some of the *least* of all *visible things*, to that *Mighty King*, that has *establisht an Empire* over the best of all *Invisible things* of this World, the *Minds* of Men." Among his observation titles are "Of the sting of the bee," "Of the feet of flies," "Of the contexture and shape of the particles of Feathers," "Of the curious texture of seaweeds," "Of the Schematisme or Texture of Cork, and of the Cells and Pores of some other such frothy Bodies," and "Of the Beard of a wilde Oat, and the use that may be made of it for exhibiting always to the Eye the temperature of the Air, as to driness and moisture." His obsession with the minutiae of this world was contagious. His drawings, which were attempting to make the invisible visible, stunned viewers.

The image of the flea especially rocked his readers, who had never before seen the intricate armor of the tiny, blood-sucking creature that had caused humans so much grief. This was the insect allegedly responsible for the bubonic plague. Perhaps readers' desire for knowledge drew them to the image—and perhaps disgust. The theorist Sianne Ngai writes in her book *Ugly Feelings* that unlike (Hobbesian) contempt—which, she argues, allows us to become disinterested and distant—disgust holds our attention. She writes that "while an object of contempt is perceived as inferior in a manner that allows it to be dismissed or ignored," the disgusting "is perceived as dangerous and contaminating and thus something to which one cannot possibly remain indifferent." Hooke's flea may have disgusted some, and the flea was certainly a near and present danger, contaminating the human body in ways that kept readers rapt, curious about the drawing of the strange armor and terrifying mouth. Hooke, however, wanted not only to shock his viewers with the flea's strength but also to show them its beauty. He wrote:

> as for the beauty of it, the Microscope manifests it to be all over adorn'd with a curiously polish'd suit of sable Armour, neatly jointed, and beset with multitudes of sharp pinns, shap'd almost like Porcupine's Quills, or bright conical Steel-bodkins; the head is on either side beautify'd with a quick and round black eye, behind each of which also appears a small cavity, in which he seems to move to and fro a certain thin film beset with many small transparent hairs, which probably may be his ears; in the forepart of his head, between the two fore-leggs, he has two small long jointed feelers, or rather smellers.

Some 350 years later, Angus is, in some ways, following in the footsteps of Robert

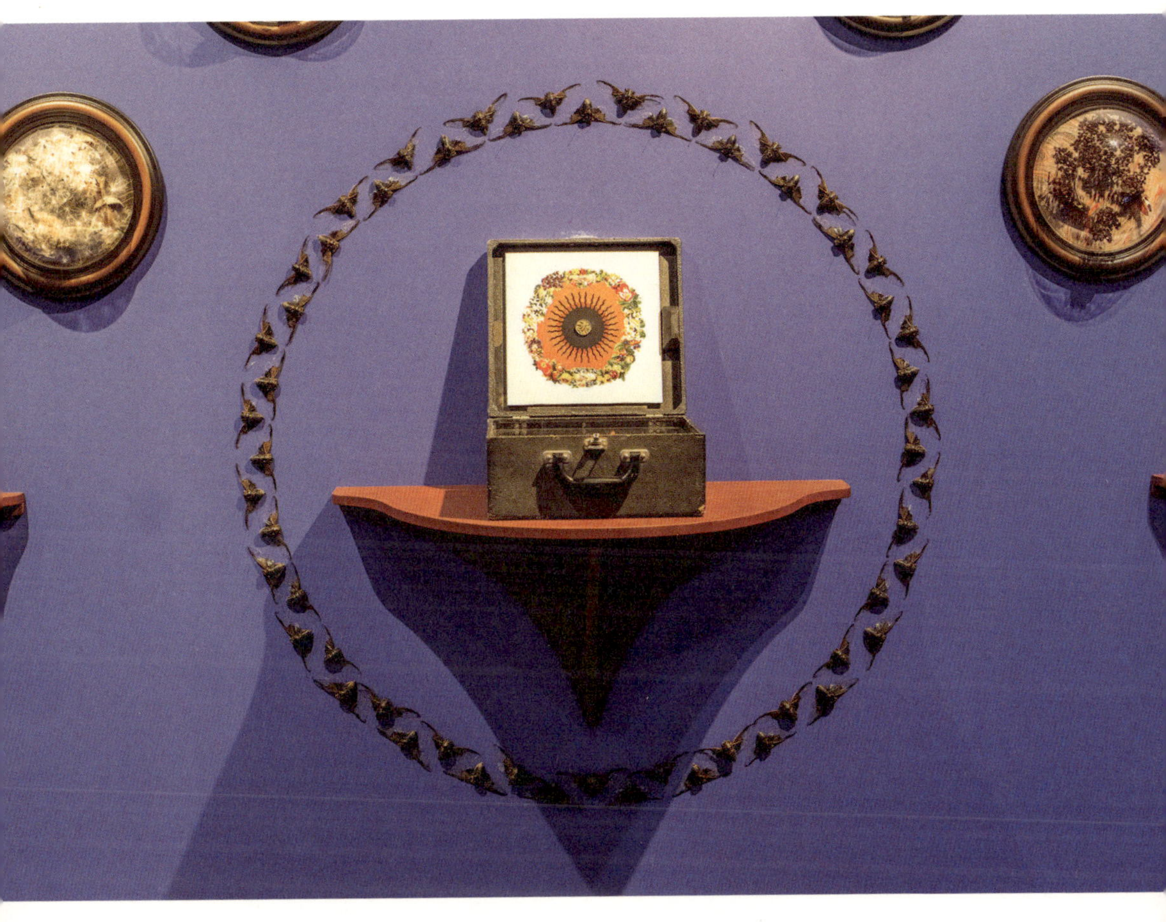

Jennifer Angus, *The Grasshopper and the Ant*. Installation view: insects, plants, view 2. Museum of Fine Arts, St. Petersburg, Florida, 2019.

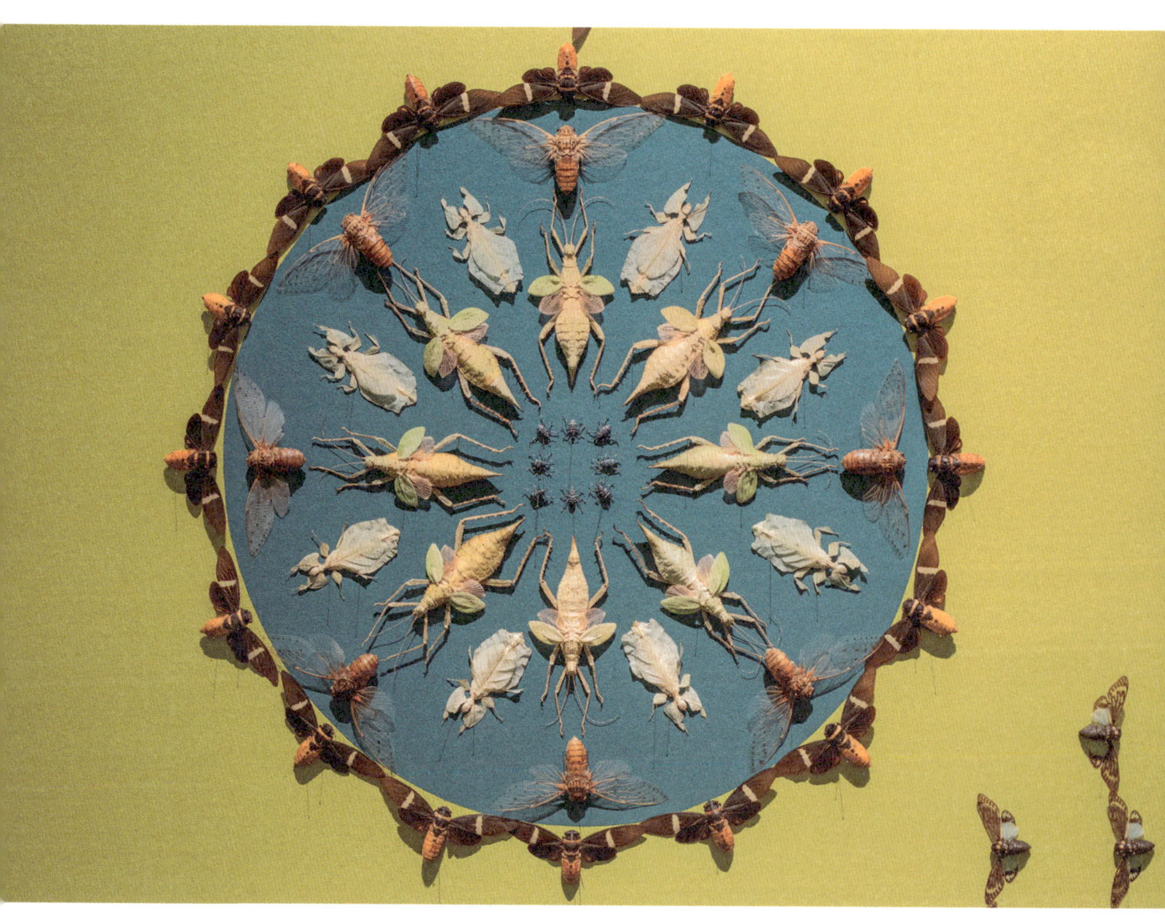

Jennifer Angus, *The Grasshopper and the Ant*. Installation view: insects, taxidermy, view 3. Museum of Fine Arts, St. Petersburg, Florida, 2019.

Hooke. Angus, who completed a BFA from Nova Scotia College of Art and Design and an MFA from the School of the Art Institute of Chicago, also wants to "offer the least visible things" to her viewers. She invoked Hooke when she talked to me about her work. At this point in history, we are quite familiar with magnification, having witnessed images ranging from viruses to plant cells and beyond. What makes insects "invisible" or "unimportant" to us now is more complicated. A large part of our blindness is born out of fear and misunderstanding. The stories we tell, the patterns we have, certainly dictate our blindness to the importance and beauty of insects. How is Angus working to change these habits? She creates exhibits that intend to unsettle the viewer. By setting up a tension between beauty and disgust, or sometimes fear, she challenges audiences to move past revulsion toward a curiosity that could lead to respect for insects—and possibly care.

In the fall of 2019, Angus had an exhibition at the Museum of Fine Arts in St. Petersburg, Florida, entitled *The Grasshopper and the Ant*. The museum invited me to give a talk about my first book during the exhibition. I was eager to see a Jennifer Angus installation in person.

The Grasshopper and the Ant began in a dark, narrow hallway. Colorful illuminated jars lined shelves on one wall, and images of stories from childhood covered the other. The wall of glowing jars looked as though it could have been an old-school pantry filled with jellies, jams, and honeys of bright orange, red, and gold. Upon closer inspection, you saw that the jars had herbs and flowers in them, and then you saw the insects. The jars were not filled with jams and jellies but rather with specimens, like those you might find in a zoological museum where insects or fish or frogs float in jars of formaldehyde. In the exhibit jars, large beetles lean up against fronds of wild mustard, a stink bug clings to a stalk of eucalyptus, a cicada lounges on an oak leaf. The effect of seeing insects floating in something that looks like jelly in a pantry is definitely unsettling. And to me, sad.

Aesop's story of the industrious ant and musical grasshopper is retold on the wall behind you. Traditionally, the story takes place in autumn, when the ant has stored food for winter and the grasshopper has not prepared properly. It is a fable meant to warn people against laziness. In Angus's retelling, the grasshopper is not prepared for winter because he has been bringing joy to the rest of the creatures with his music—a subtle defense of the artist and, perhaps, the insect.

At the end of the hallway is a door that leads to a bright room with chartreuse walls covered in intricate designs reminiscent of Victorian wallpaper or embroidery. I could hear audible gasps as people realized that lines of cicadas, grasshoppers, and beetles, all pinned to the wall, made the thick, vine-like patterns. In the center of the room was a large table set with an elegant dinner, surrounded by life-size animals. It's a little like walking into a scene from a children's book.

Children's stories are often filled with nonhuman characters who have human characteristics. In Aesop's fables, the animal stories intend to teach their young readers about morals. The hare is famously overconfident during a race against the tortoise—so confident that he falls asleep. The tortoise keeps going and eventually wins not by speed but by persistence. A mouse whose life was spared by a lion frees the lion from a net, reminding the child that no act of kindness is wasted.

In E. B. White's *Charlotte's Web*, a spider writes words like "radiant" in her web to save a pig named Wilbur. In *The Cricket in Times Square*, a country cricket learns to play classical music in a New York subway, with the assistance of a cat and a mouse, to help out their friend, a boy whose parents own a failing newspaper stand. In Winnie the Pooh stories, a bear, a pig, an owl, a tiger, a kangaroo, and a boy have adventures with bees, honey, and heavy rain. All of these stories seem plausible, and the characters feel real to the open mind and heart of a child.

It all seems very cute, because as adults in the modernized Western world, we learn that all of those connections with animals are imagined. Animals can't actually speak our language. Animals, according to Descartes, were merely machines. The prevailing story has been that animals are not as smart as humans, not as capable of problem solving or of having feelings. Science, however, has proven this to be false on countless occasions. Octopuses solve problems, crows use tools, elephants mourn, honeybees experience stress. But in general, insects are largely disposable, if not outright enemies.

Many children who are connected to animals when they are small, who think of them as equals or at least worthy of respect, lose this as they age. Insects, especially, become enemies. Many insects become pests. As adults, our relegation of insects to the realm of pests or something to be feared and killed creates devastating patterns on our planet.

Patterns can be beautiful and still be destructive. When I saw the circular patterns on the wall, I thought of looking down from an airplane at the quilt of industrialized agricultural spaces we have created in the United States: miles and miles of monocrops, where pesticides work to destroy insects of all kinds and rich ecosystems have been turned into ecological wastelands. Or the patches of lawn that require chemical treatments every three weeks to keep anything from growing besides a carpet of one species of grass. No habitat for insects or birds or amphibians. These are the patterns we make as adults. Only recently, as scientists have begun to show evidence of dramatic declines in nonhuman species, have we started to become aware that perhaps our habits, our patterns, and our stories need to be examined and changed if we want the creatures with whom we coexist to survive.

Angus is keenly aware of the danger insects are in—and the danger we are in as a society. One of the most frustrating comments she received after a show was one that accused her work of being "insecticide." Angus does have a huge collection of thousands of dead insects, which might be seen as being part of the problem and not the solution, but she is quick to point out that all of her insects are responsibly harvested and that after a show, all of the insects are

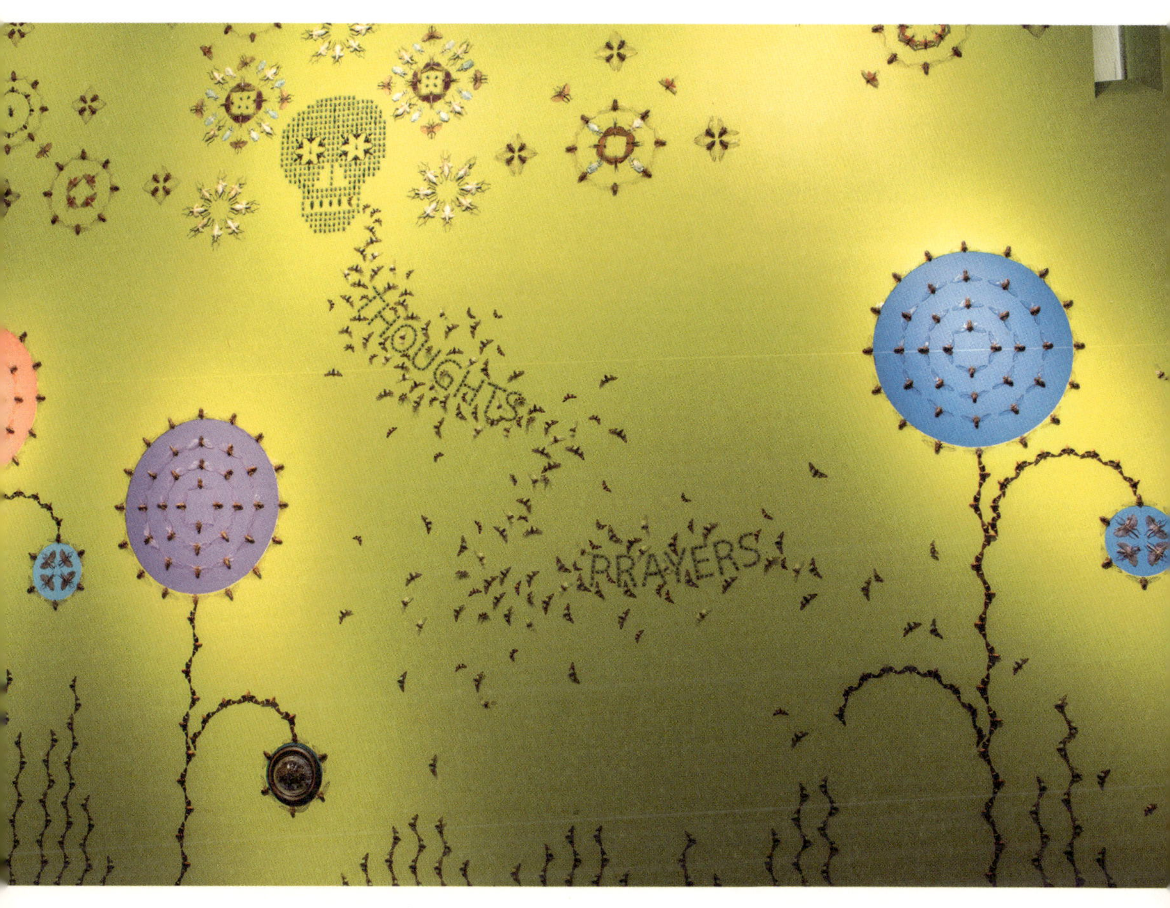

Jennifer Angus, *The Grasshopper and the Ant*. Installation view: insects, taxidermy, view 4. Museum of Fine Arts, St. Petersburg, Florida, 2019.

taken down and carefully stored in boxes for the next exhibition. She's had these same insects for years.

At her show, I spent a long time looking at a group of pink-winged grasshoppers, each easily as tall as my palm. This grasshopper's lower two wings are like pink fans, flung out from its thorax above its powerful long legs. They were beautiful. A quick Google search revealed that I could buy some in bulk for fifteen to thirty-five dollars; the advertiser claimed that the insects were sustainably harvested in Madagascar. Other people clearly thought they were beautiful, too. I didn't buy any myself.

Jennifer Angus shares her insect collections to make people think. She wants us to question our preconceptions, our feelings of disgust or repulsion, and to examine our stories, to consider our patterns, to rediscover the wonder and respect for insects we had as children. And just maybe to discover that insects deserve our care. I was determined to learn more about how people were working to keep our ecosystems and our living insect populations intact.

Chocolate and Coffee

SAVING WHAT IS BEING LOST

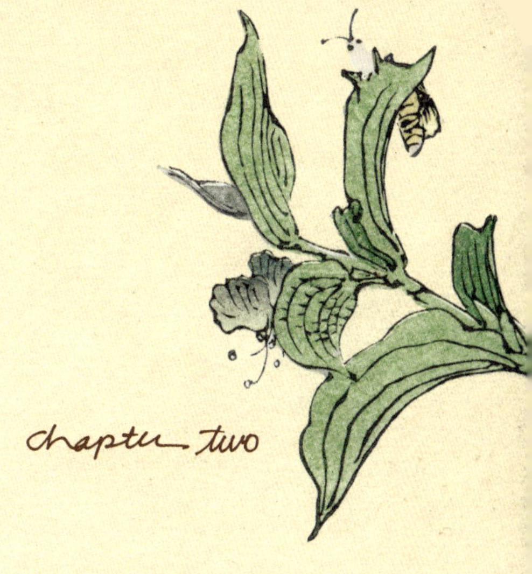

chapter two

Early morning mist billowed across the tops of the surrounding blue-green mountains as our rental car bounced up a rut-covered dirt road. Over coffee earlier, a local resident had suggested that there would be "birds near the waterfall," and so we were headed in that direction. Sunlight streamed through the moss-covered trees, dappling the wet leaves below. My traveling companion, Drew, and I were looking for a place to hike. I had come to Ecuador to see what "rich biodiversity" looked like firsthand and to meet some farmers who were interested in trying to preserve places like this by thinking differently about farming.

After twenty minutes of a painstakingly slow ascent, a small hand-painted sign caught our attention. "Tanager Reserve," it promised. There was no obvious trail or place to park, but we pulled over to the edge of the road, parked the car, and stepped into the moist air. The faded wooden sign said that the reserve was only two kilometers away and suggested leaving a five-dollar donation. A very weathered image of a bird roughly resembling a scarlet tanager added to the sign's sincerity. But we could see no trail. Farther up the road? Then we saw a very narrow divot in the forest's edge, like hair parted with a comb. On closer inspection, we saw that it was indeed a path, but it looked more like a tiny gulch carved by a downpour during the rainy season—muddy and only as wide as my foot. Plants and trees leaned in from both sides. My very American idea of a trail was being challenged. Could this really be it? We stood deliberating. Heavy rain was in the forecast, and we could see that if there was indeed a trail, it would certainly wind down the side of this mountain and would, we realized, potentially be dangerous because of the very common mudslides.

At that moment, three men dressed in khakis and field hats emerged from the dense foliage. These were the people, we learned, who maintained and did research at "the lodge" on the Tanager Reserve. They said they were going out for a while but would return eventually. They encouraged us to check out the trail, then turned and headed down the dirt road on foot. Without another thought we plunged into the dense greenery. Within twenty yards, the trail opened out slightly to a muddy footpath, now as wide as three of my feet side by side. The air rang with the sound of creatures of all kinds calling out to one another. Every leaf glistened and seemed to quiver to the verdant song that was the cloud forest, the *bosque nuboso*. We knew this would be a good spot to see birds and insects.

About an hour north of Quito, Ecuador, in the middle of the cloud forest, this little town called Mindo, where we had had our morning coffee, draws huge numbers of a very specific type of international visitor every year: the birder. Both professional ornithologists and amateur birders flock to this region because it is rich in birds. For years, Mindo has been a prime spot for birders competing in the Audubon Christmas Bird Count. An ornithologist named Frank Chapman initiated the count in 1900 because he wanted to shift an annual tradition of competitive bird hunting to a competition for seeing, naming, and recording species. His plan succeeded. The annual count currently provides invaluable data about the fluctuations of bird populations through time. Every year, birders all over North and South America get out their binoculars and notebooks and go into the field to record what they can see and hear. For many years, Mindo has recorded one of the highest species counts and is therefore one of the most exciting places to bird. In 2017, during the 117th annual Christmas count, Mindo topped the charts with 456 species recorded. In 2018, Mindo was second only to another biodiversity hotspot in Ecuador: Yanayacu, in Napo Province, which recorded 502 species. The whole country is home to over 1,000 species. In the United States, the highest bird counts in 2018 were in Texas, at the Mad Island Marsh, where birders found 220 species. The superstars of the Ecuadorian bird world include the Andean cock-of-the-rock, the yellow-throated toucan, the giant antpitta, the golden-headed quetzal, the scaled fruiteater, and the torrent duck. Clearly, Mindo would be attractive to serious bird people. But Mindo is not only attractive to birders; it draws nature lovers of every ilk. As a person concerned about insect populations and biodiversity, I wanted to visit a place where a truly rich ecosystem, which included plants, insects, animals, frogs, birds, and mammals, still thrived. I made a plan to go to Ecuador.

During the weeks I was planning for my trip, the United Nations published a riveting biodiversity report. Across the globe, newspapers suddenly bore shocking headlines about the nonhuman world. "A Million Species at Risk of Extinction," they said. The UN report, produced by compiling the studies of interdisciplinary experts around the world, issues a dire warning to the citizens of the earth. We need to shift our practices because ecosystems are collapsing. Regardless of whether one cares about the animals and insects involved, given the fact that humans depend on ecosystem health for food, this report should make our hair stand up on end: "The average abundance of native species in most major land-based habitats has fallen by at least 20 percent, mostly since 1900. More than 40 percent of amphibian species, almost 33 percent of reef-forming corals and more than a third of all marine mammals are threatened. The picture is less clear for insect species, but available evidence supports a tentative estimate of 10 percent being threatened." These declines are unprecedented, and many of the insects we count on are in trouble. The study out of Germany in the fall of 2017, which I mentioned in the introduction to this book, reported a decline of 75 percent of flying insects over a twenty-five-year period. Pollinator populations, including those of many kinds of bees, butterflies, and others, have been shrinking in number in many studies. Pesticide use, habitat destruction, and climate change all affect their health. Their decline will result in a decrease in food diversity and availability for humans and nonhumans alike.

One of the co-chairs of the UN report, Professor Josef Settele of Germany, said, "Ecosystems, species, wild populations, local varieties and breeds of domesticated plants and animals are shrinking, deteriorating or vanishing. The essential, interconnected web of life on Earth is getting smaller and increasingly frayed." Settele added, "This loss is a direct result of human activity and constitutes a direct threat to human well-being in all regions of the world."

The report was devastating to read. But while my heart was heavy, my sense of urgency to tell stories about our planet was bolstered. What is biodiversity, anyway? Could any of us recognize it?

As many know, Ecuador is a country sitting on the equator, a country boasting outrageously diverse landscapes, including the Andes, the Amazon, the Galapagos (Darwin's stomping grounds), cloud forest, and coastal lowlands. It is the home of my friend Liz Hennessey's beloved tortoises. A country once dominated by eight Indigenous groups, it was eventually taken over by Spain, and today a residual mix of nationalities and languages, including expats from Germany, France, and the United States, make up the population.

Ecuador is also one of the most intact biodiversity hotspots in the world—"more biodiversity per square kilometer than any other nation on earth," according to *Newsweek*—and, very importantly, one of the first countries to recognize the rights of nature. The constitution of 2008 includes a tenet called Sumac Kawsay, which is Quechua for "life at its fullest." It "guarantees the right to live in peace and free of contamination. It grants the environment the same rights as a citizen, including the rights to be respected in its entirety through its vital cycles, as well as the right to restoration. This tenet of the constitution also promises to involve and consult all local populations about activities that have an impact on their environment." Upholding this tenet can prove difficult, I would learn, but these values and spaces have made Ecuador's ecotourism industry very successful.

Ecuador also produces very good chocolate.

I had been drawn to Ecuador originally because I was thinking about the importance of insects to ecosystem health—specifically, pollinators that coevolved with certain flowers and plants. Ecuador had lots of examples of this beautiful symbiosis. Cacao plants and their insect partners, species of *Forcipomyia* or chocolate midges, provide one powerful example of an important insect/plant partnership. The chocolate plant is entirely dependent upon one insect for pollination. A fascinating book by Erica McAlister, *The Secret Life of Flies*, fully examines the stories of our uncelebrated insect companions, and one section is dedicated to the midge. Without this specific type of midge, there would be no chocolate. While chocolate was thought to have been cultivated first in Mesopotamia, recent studies published in the journal *Nature Ecology and Evolution*, using "cacao starch grains, absorbed theobromine residues and ancient DNA—dating from approximately 5,300 years ago recovered from the Santa Ana-La Florida (SALF) site in southeast Ecuador," suggest pre-Columbian use of the plant. The human relationship with chocolate in Ecuador is thousands of years long, but we don't often talk about our nonhuman conspirators, the midges.

The flowers of the cacao plant are incredibly fussy and so small, at one to two centimeters, that most other pollinating insects will not be able to access the delicate parts important for pollination. The midge, however, at two to three millimeters, is small enough and very, very hairy—its antennae look like witches' brooms!—which is good for these finicky flowers. The midge is also good at navigating the complex anatomy of the somewhat chaste chocolate flowers. Strangely, the tallest, most erect parts of the flower, the staminodes, do not produce pollen. The pollen-producing male parts of the flower, the anthers, are hidden by hoods that must be uncovered by the midges. The styles, the female parts, which want the pollen, are located in the middle of the tower of staminodes. The midge is the only insect who has successfully evolved to satisfy this coy plant. It thrives in a very specific climate, one with lots of shade and moist soil. As we face a warming climate and radical shifts in ecosystems due to industry and agriculture, I wondered, are these insects at risk? How can we keep important insects healthy? Ecosystems thriving? And economies dependent on these ecosystems strong? How could we preserve these extraordinary relationships? I wanted to talk to cacao farmers. Ecuador was a good spot to do this. I knew that industrial agriculture was a threat. What I didn't know was that this precious, bio-rich region was also in danger of being destroyed by other means: mining.

Although we are not true birders, my friend Drew and I were keen on seeing as many birds as we could. At a little restaurant catering to international travelers called the Beehive, we had scanned bird books over *cortados* sweetened with the local honey. While many seek out the most exotic, like the cock-of-the-rock and the toucan, we would seek out some of the twenty types of tanagers, favorites of Drew's. Knowing that tanagers (and many other birds) are primarily insect-eaters, I knew that there must be an incredible diversity of insects out there as well.

I was living in the Midwest region of the United States, most of which had been flattened by glaciers out into wide plains, a place once filled with bison and inhabited by groups like the Ojibwe, the Menominee, and the Sauk. It had been cultivated for centuries, with the dominant crops in the last fifty years being very hybridized and chemically dependent corn and soybeans. The biodiversity in those spaces is next to nothing.

That morning in Mindo, even though we were not going on a formal tour, I assumed that we would see a range of animals, plants, and insects I had never seen before. I was not disappointed, but I also realized why people hired expert birding guides to take them through the forest. As we headed toward the Tanager Reserve, the path seemed deceptively like a tunnel because the foliage on both sides created a tube of green—but one misstep and a body could plunge down the slope of a deep ravine. The branches to our right were treetops covered with epiphytes. Rivulets occasionally trickled down from the slope that extended up to our left, carving away at the already precarious trail. The required pace, one that kept us safe and fully present, allowed us to notice the life around us, an attention not often used in our high-speed technological moment. The forest lived, breathed, pulsed around us. Mosses covered stones and roots. Epiphytic bromeliads grew impossibly out of branches without soil or any obvious water source. We saw leaves the size of a sparrow's eye and palm leaves large enough to nap on. Vines coiled up and around trees; tendrils of baby ferns unfurled their magnificent spirals. So many shades of green punctuated by bits of brightness: tiny red blossoms bright as the most exciting nail polish, delicate fronds hanging overhead ending in pouchlike pale pink flowers whose openings looked like the orchids I saw in the grocery store at home. So many plants. I could never identify them all. And then there was the sound. This was by no means a quiet forest. Bird and insect voices resounded in the air, the cloud forest chorus.

Although the many creature sounds engulfed us, seeing individual birds and insects proved more difficult than we expected. We moved cautiously around the edge of the mountain, our eyes scanning back and forth between the path and the canopy. Butterflies and bees floated around us. One particular butterfly, with transparent wings edged in black and pink—a pink-tipped clearwing satyr—caught my eye, and up above, a wide range of birds moved too fast for us to identify them. Once, a larger turquoise-winged bird flew up suddenly out of hiding. We glimpsed its red belly and long black tail. A bird identification book later told us it was an Ecuadorian trogon. We kept walking toward the reserve, an occasional four-by-five-inch sign assuring us that we were still, in fact, headed to a destination of sorts. At one point, the path disappeared; it had been replaced with a wide swath of red mud, a small mudslide. From the fresh boot-prints, we surmised that the new mud had been traversed by the researchers we had seen earlier. We crept across, hoping the soil would continue to hold, and luckily it did. At a place where a stream crossed the path, we encountered dragonflies of several kinds along with a giant red millipede, as wide as my index finger and twice as long.

The roar of rushing water began to drown out the bird and insect song. At the base of the gorge we had been climbing into, a riotous river surged over boulders and stones. "Tanager Reserve Ahead," a tiny sign whispered. Soon we came to a bridge. Two thin cables connected trees on both sides of the river. Suspended from the cable by ropes were wooden slats. Here was a different sign: *Aviso! Una Persona a la Vez! Caution! One Person at a Time*. I went first, the lighter of the two of us, holding onto the shoulder-high cables for balance as the wood slats swayed and bounced beneath my feet. The gushing water sped past, capable of swallowing me whole. For a moment, as the beautiful and slightly terrifying power of that water humbled me, I understood the old notion of the romantic sublime—a feeling admittedly rare in this historical moment, when humans are so powerful that we have dubbed this age the Anthropocene. I walked cautiously across. Drew followed once I reached the other bank.

"Tanager Reserve," one more faded, hand-painted sign announced. No gate. No fence. Just more trail. The only notable evidence of the reserve was a small wooden building up a hill among the trees and vines. In no time, it seemed the forest could simply swallow the building whole. Here we were, at the destination. We looked around for tanagers, but to no avail. As we walked uphill, the sound of the river was replaced with a new sound: the sound of hummingbirds whirring through the air, a sound not unlike the sound a lightsaber makes as it cuts through the air in the *Star Wars* films.

Hummingbirds come in many sizes, and they can beat their wings from fifteen to eighty-eight times per second and reach speeds of fifty mph. The ancestor of these specialized nectarivores is said to have lived in South America twenty-two million years ago. No one was in the building, so we sat on the tiny deck and watched hummingbirds come to some feeders placed there. There must have been at least four or five at the several feeders at all times, sharing the sweet liquid with bees. So many different kinds! Iridescent green wings, black wings, brown wings and white bellies. Several had what looked like fuzzy white boots, and one had a tail twice as long as her body, reminiscent of a quetzal, which seemed to make no evolutionary sense. And to my memory, Darwin made evolution seem so logical. These birds may not have received the memo. They did share the long beaks so good for reaching deeply into specific flowers. Many kinds of bees also have extremely long tongues, good for plants with deep nectaries. Perfect companions for the wide range of flowers we saw blooming on the trail. But I couldn't begin to identify them all. I saw how hard it would be to perceive what was missing here or which partnerships were most important. The complexity filled me with awe.

I knew that all around us were larger animals unwilling to show themselves. Among those were the coati, the kinkajou, capuchin monkeys, and armadillos. This was what biodiversity looked like, I was thinking. This is what the UN report warned was in danger. But what partnerships kept this and other specific ecosystems going? Which birds relied on which insects? Which insects relied upon which plants? This is why people like E. O. Wilson urgently and idealistically called for a "half-earth" strategy of conservation, allowing for the preservation of intact ecosystems. The UN report was devastating, but could *all* of this be at risk? In fact, I later learned, this very spot had been slated for mining.

While Ecuador is heralded as a biodiversity hotspot, with some areas even designated World Heritage Sites, the truth is that much of this land was preserved when the country was enjoying an oil boom. During the days when President Rafael Correa was in power, the fossil fuel economy allowed for programs and policies—including the constitution that supported the rights of nature—that encouraged keeping forests intact and keeping wildlife areas in conservation status (as well as building new roads). But in lean economic times, when the country needed to boost its revenue, industries threatened these essential spaces of biodiversity. Huge loans obtained from the international community put pressure on the Ecuadorian government to open up the forests for mining. The government has rights to everything under the ground, which in Ecuador means gold, copper, and silver. Were these a gift or curse? Some have celebrated this economic resource. In 2019, *Reuters* announced that "two of Ecuador's five mines in development are on track to start producing copper and gold in the fourth quarter of 2019 in line with plans, a senior government official said in an interview, as the country pushes to diversify its economy from oil exports. . . . The OPEC nation, which has struggled with low oil prices in recent years, has been drumming up foreign investment to tap its big copper, gold, and silver deposits. It aims to more than double the value of mining to its cash-strapped economy by 2021." Others adamantly protest these plans because they care very deeply about the ecological riches, ecotourism dollars, and Indigenous territories that will be lost. A website for the organization The Rainforest Action Group has published maps of proposed and current mining treaties. There are dozens in protected rainforest areas. Including the space we were standing in.

At the top of the trail, when we emerged from the forest, we noticed three people not far from where we had left our car. One of them had a laser pointer, and the other two were silently gesturing and pointing with excitement when his light landed on a branch. A professional birding guide, we realized. Drew walked over toward them, just in time to see a golden tanager. His lucky day!

On the road back to the town of Mindo, I thought about how hard it was to convey the value of a place like this, even perhaps to someone standing right next to me. We live in a world that finds it difficult to assign a value to a forest, much less an insect or a plant or a bird. Ecotourism was one solution. People from all over the world traveled here to see these special places. But it is a delicate balance to strike. One needs only to read Jamaica Kincaid's *A Small Place* or the much more recent critique of tourism in the Himalayas by Van Badham to be reminded that tourism can be very problematic. But done well, perhaps it can be good for locals and for the land. It certainly seemed better than strip mining.

Before we left Mindo, we stopped at a butterfly sanctuary. Here was a rare example of an insect drawing a public audience and an income for locals. A wooden gate opened to a garden and a small shop in front of a long, low building with a tin roof and screen walls. We made our way through a curtained doorway into a narrow hallway, another curtain, and then a room filled with flowering plants and butterflies. The air itself seemed alive, fluttering. Quickly the electric blue of the blue morpho caught my eye. The guests walked slowly and carefully, eyes wandering in astonishment, some spellbound by a winged creature pausing on a finger or a sleeve. Wide-eyed children watched as owl butterflies feasted on open blossoms or a tray of ripe bananas. I paused, captured by the slow unfolding of the *Heliconius clysonymus*'s wings, like hands moving from a position of prayer to offering. I tried imposing my limited knowledge of North American butterflies to the other beauties, to no avail. Ecuador was home to some striking butterflies, to be sure, including the variable cracker (*Hamadryas feronia*), which looked like a pewter blue-and-gray tapestry, and the neglected eighty-eight (*Diaethria neglecta*), whose black-and-white designs look like they were painted by Keith Haring. I thought then of Samuel Green's poem "Butterflies," about a woman who is passionate about identifying the ones she finds in her yard, "letting her sweet, light attention land / on one luminous thing after another." Here we all were, taking the time to let our attention land on all these luminous things.

One wall was devoted to chrysalises. Some looked like golden autumn leaves neatly sewn up one side with tiny brown stitches, others like tiny pots dipped in a celadon glaze my mother might use, and still others rough and brown like crumpled bark. One honestly looked like it had been dipped in gold. In the older cocoons, you could sometimes make out the indistinct shape and color of a wing. Astonishing. I made a mental note to learn more about the transformation happening within these special vessels. It was definitely worth the entrance fee. Could butterfly tourism help save this place from mining? And what about the rights of nature?

Was there hope for protecting the home of the trogon, the centipedes, and the midges? Experts in law like Mary E. Whittenmore have doubted the power of the "rights of nature" constitutional amendments and have argued that "plaintiffs who sue under the amendments face significant legal barriers, such as Ecuador's lack of a standing doctrine and a history of judicial corruption and dysfunction." But in 2021, the government of Canton, home to forty-three Indigenous communities, challenged ENAMI, a mining company with plans to mine . . . and won in Ecuador's highest court. The land, which was home to 178 threatened or near-threatened species, would be protected. The case made ripples across the globe. "Policy frameworks that place humans in context as a part of nature . . . rather than placing humans as above, or apart from, nature, will be a necessary part of addressing the serious environmental issues that our planet is facing," said Mika Peck, a senior lecturer in biology at the University of Sussex.

Mining was certainly not the only reason we were losing biodiverse spaces like this. Industrial agriculture was another. I knew there were farmers in Ecuador who were thinking hard about how to grow food for people while also respecting the ecosystems they lived in. We would head to the coastal lowlands to meet a few of these farmers.

Leaving the cloud forest behind, we drove to the coast. As the elevation changed, forests became less dense and the landscape less vertical. We passed roadside stands selling bananas, mangoes, and coconuts. Small towns made up of one-story cement block buildings with metal garage doors that opened up to stores offering everything from platano chips and brooms to auto mechanic services. We also passed hillsides that had been scraped clean of any growing thing. Some fields were still strewn with dead trees, which struck me suddenly as so much like fallen bodies after a horrible disaster. Beyond these were billboards promising Palm Paradise, and then we saw the vast acres of the palm plantations. This was a landscape I knew well from driving past miles of cornfields in the Midwest. Nothing between the rows. Michael Pollan has said that in the natural world, monocrops are senseless. Palm plantations like these were popping up all over the globe. And I thought of how often I didn't check ingredient lists for palm oil. I was part of the reason these were here. A sinking feeling, familiar these days, filled my body. I wanted some hope. I couldn't wait to meet the farmers.

I had learned about this particular group of farmers, held together by a pledge to keep their land intact, from Joe Meisel and Catherine Woodward, who led an organization called Ceiba. The Ceiba Foundation was committed to preserving landscapes with the cooperation of Indigenous people in Ecuador. Ernesto Campo and Vicko Castro both farmed land their families had owned for many years. Ceiba supported them by getting them involved in a forest corridor rehabilitation project.

A river over the road to Ernesto Campo's place made it impossible to reach by car. We crossed a makeshift bridge and walked for an hour in the hot late-morning sun toward the small farm. Still drying out from the rains the night before, damp patches—left behind as the puddles evaporated—covered the road. These patches attracted butterflies of every size and shape. Clouds of fluttering wings lifted into the air as we approached. I remember yellow sulfur butterflies gathered on a gravel road near the house I lived in as a very little girl. These days the butterflies are scant there. But here, I saw not only butter-yellow butterflies but an array of colors! Pewter blue, sky blue, bright yellow, white, orange, dark brown . . . ! For me, the scene far surpassed the butterfly display in Mindo.

A simple arbor made of logs draped in magenta bougainvillea marked the entrance to Ernesto's place. We walked down through a corridor of trees and flowering bushes to a clearing where he'd built his house, a simple wooden structure perched ten feet off the ground on thick posts, with no windows, just walls that ended about three feet from the floor. He and his daughter Lourdes walked down some wide wooden steps to greet us, smiling broadly. Ernesto's obvious vitality belied his eighty-four years. He ushered us in with warm words and vigorous full-arm gestures. We sat on the porch of the house as they told us about their farm and the reasons for their organic agroecology methods. The farm did not conform to the picture I had in my mind of farms from my own history. There were no rows, no obvious division of species or types of tree or plant. And yet, he explained in Spanish, his farm produced guava, coconut, bananas of many kinds, plantains, limes, passion fruit, pawpaws, oranges, mangoes, coffee, and chocolate. Behind the house, as some bananas—still in their peels—cooked over a fire, Ernesto showed us some trays of cacao beans, which require "five times of sun" to dry out before they could be ground into the bitter cacao we would recognize. A few chickens poked the earth nearby for bugs. Birds called out from the trees. Lourdes smiled knowingly as she watched me take it all in.

Lourdes and Ernesto had grown accustomed to visits, because the Ceiba Foundation had enlisted their cooperation in being a model farm for other farmers nearby. Ceiba was working with a group of farmers along the Ecuadorian coast, which is home to a unique ecosystem called a tropical dry forest. The integrity of a forest like this is disrupted by logging and traditional agriculture. The agroecology model allows the Indigenous and local people to continue to obtain food from the forest while keeping forests intact. The corridor project was essentially maintaining and rejoining forest lands so that habitats and soils would remain stable. Ceiba was "promoting forest protection, conservation incentives for land owners, land purchase, and locally-managed reforestation projects, to connect over a quarter-million acres of forest fragments through a groundbreaking Conservation and Sustainable Use Area (ACUS, in Spanish)." The foundation aimed to create "a biological corridor spanning over 135 km (85 mi) and a range of micro-climates and habitat types." As the climate grew drier, it would be more and more important to keep areas forested to provide shade and soil integrity. I thought of the chocolate midges, so dependent upon a specific temperature and necessary to pollinate the cacao plants. What would life be like without their survival? My own diet would certainly change quite radically. Ernesto and Lourdes used farming methods that other landowners could emulate. Could they also be models for farmers in my own region of the United States?

Lourdes walked us through the "farm," which was essentially a forest. We walked past all kinds of fruit trees—banana, mango, pawpaw—until she stopped at one. This was the type of tree that supplied me with dark chocolate, hot cocoa, flourless chocolate cake, and mole sauce. She wielded the machete she'd kept at her side with a decisive chop and sliced off a thick pod. Cut open, it was filled with oval seeds the size of quarters covered in a thick, milky substance. She directed us to put them in our mouths and suck them. The taste was sweet and subtle and tasted nothing like chocolate to me. These were the seeds that Ernesto would put in trays to dry in the sun. Next she called her son. He approached with a long pole and knocked fresh coconuts from the top of a tree. With his own machete, he chopped the tops off of coconuts the size of small bowling balls and encouraged us to drink. It may have been the heat of the long walk, but that fresh coconut water refreshed and nourished me like few things in my memory.

Before we left, I bought a bag of cacao, while Ernesto told me about his plans for his next birthday, which would include lots of dancing. His great health, he claimed, was from eating organic food and refusing to eat red meat. His huge grin and merry eyes were convincing enough for me.

Before visiting another farmer, we stopped at Ceiba's Lalo Loor forest education center. Kelly van Gils, a forestry conservationist from the Netherlands, met us at the main building. The center rested on two hundred hectares that had once been damaged by overgrazing livestock and illegal hunting. The area was rich in guinea pigs (an Ecuadorian delicacy) and howler monkeys, both attractive to hunters. Twenty-five years earlier, the Lalo family had decided to enclose the land and restore it to its original forested state. Ceiba teamed up with them to help them make it into an educational space for local schools and tourists. They made trails and helped expand the forest. Ceiba, van Gils explained, worked with the local library and schools to bring children into the forest to learn of its value and to participate, hands on, in the reforestation projects. College students from around the world also came to visit the forest. Native trees and plants were planted on the property, including orange, lemon, avocado, mandarin, and coffee plants. The land was home to 190 species of birds and 40 mammals—and, recently, 20 boxes of honeybees to help with pollination. Two howler monkeys played in the treetops as van Gils spoke. I was distracted, as I had never seen any in the wild. Educating the community, especially youth, about the value of this land and these unique creatures was crucial.

We hopped back into our rental car and drove toward Vicko Castro's farm. At a certain point, once again, we had to abandon our vehicle and walk over washed-out roads to get to her land. Castro wore a long black dress that matched her hair and eyes, and though her gaze and her posture were even and strong, and maybe even a bit intimidating, her voice and smile welcomed us in. She and her partner, Daniel Recalde, stood near their home, which was something like a tree house erected at the bottom of one of the surrounding hills. They urged us to come in for a cup of coffee before we walked her land—coffee she had grown, right on this property, that was roasted by a friend in the nearby town. We sat at long wooden benches, and she poured us each a cup of the strong dark liquid that smelled like the rich earth around us. It was a perfect cup of coffee.

Castro was part of a cooperative of coffee growers consisting of twenty-four families and eight different communities. They shared a commitment to ecological practices and the desire to make a sustainable income for the Indigenous people who lived in this region of Ecuador. Their cooperative afforded them the opportunity to pull together resources for roasting equipment, packing materials, marketing, and distribution. They also had workshops to share ideas about soil improvement and pest management using alternatives to harmful pesticides. Conservation starts in communities, she said. It improved well-being for the farmers and for the land.

After coffee, Castro led us on a tour. Her family had owned this land for years. As a young girl, she yearned to see the city, and when she was old enough, she moved away for education—but her love of the land brought her back. She couldn't stand the idea of being in an office every day, she said, when she could farm instead, so she came home. With the help of several other people who were living on the farm, she transformed the land into an organic, shade-grown coffee farm. The coffee plants coexisted with the myriad other species. Castro practices growing coffee in the shade, as it is better for soil and allows for more biodiverse landscapes. Plantains supply the shade for the coffee plants, primarily, but we also walked past guava, lime, mango, and passion fruit trees. Reforestation took years, but the hillside was now lush with a wealth of plants and trees.

On the way back, I asked Castro which bees pollinated the coffee. She grew very excited and ushered me to a pillar near her house. There she had me stand on a chair so I could get close enough to see the little native bees that had made a nest there. I watched them landing on their tiny delicate feet, all walking toward a small circular hole. So much smaller than the honeybees I was so familiar with. Stingless native bees, Castro whispered excitedly. Sloths, apparently, found them tasty. I knew Ecuador was home to at least 132 species of stingless bees. Many of them were at risk as a result of development and climate change, but good practices, like growing coffee in the shade, could help keep the bees and the coffee thriving. We watched for a moment, reverent. I marveled at their quiet industry. Then she took my arm and led me to a wasp's nest. These wasps were eating pest insects, she told me.

I was rather giddy. A kaleidoscope of living beings all working together—it could be this way. All of this richness coexisting. Knowing what I did about pollinator decline, the problems with pesticides and monocropping, the mining threats, the threats to the lands of the Indigenous people, and climate change, I saw that Vicko Castro and Ernesto Campo were doing the important work, the work that could truly help the planet. I told Castro that she and her farm represented real hope, in my opinion. She beamed and nodded slowly, with the gratitude and humility of a sage.

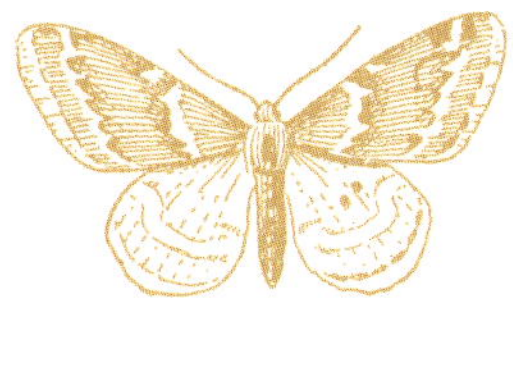

Lea Bradovich

Metamorphosis

Metamorphosis—noun: 1. (in an insect or amphibian) the process of transformation from an immature form to an adult form in two or more distinct stages. 2. a change of the form or nature of a thing or person into a completely different one, by natural or supernatural means.
—LEXICO

The life of a butterfly defies logic. These insects deftly transform themselves from something cute, wormlike, and earthbound into something elegant, intricate, and winged. They are shapeshifters, capable of alchemy, of the magical process of metamorphosis. The butterfly begins as an egg, becomes a caterpillar gorging on leaves until she is prepared to form a chrysalis and finally to emerge as a colorful creature capable of flight—a butterfly. How is that possible?! Eric Carle's *The Very Hungry Caterpillar* seals this story into the minds of many children. The process is such a profound symbol of the capacity to change, to grow, to transform, and thus the butterfly is a ubiquitous image signifying the potential for, or sometimes the promise of, renewal. This change is not a simple process. Here is the poet James Crews, imagining what the transformation might be like for the insect, and metaphorically, for ourselves.

MONARCH

The butterfly does not break free triumphant.
Once it claws through the chrysalis,
it stands there shivering, new wings aching
as they slowly fill with blood. It must keep
its tiny eyes shut tight at first against
the brightness and shimmer of a world
it has never seen before—not like this.
It must listen until a deeper voice whispers:
The flowers are waiting. Leave the skin
of the old life far behind. Open your eyes
and give in to the blue air that will carry you
everywhere you need to go.

In Lea Bradovich's painting *Metamorphosis Dress*, all stages of this process are on display. The painting is of a woman whose dress is made of emerald milkweed leaves, the essential food for monarch butterflies. Monarch caterpillars curl around her head, chrysalises in all stages of becoming hang from her chest, and adult butterflies cluster by her face. Her expression seems pensive.

I have a vivid memory of visiting the Mindo butterfly sanctuary in Ecuador, one home of the dazzling electric-blue morpho butterfly. The pulse and float of the insects in flight made the light in the room flicker like an old silent film, but in color . . . butterfly wings with patterns of orange and yellow, blue and black, red and brown flew around us. It was there I saw my first live *Caligo atreus*, or yellow-edged giant owl butterfly, whose wingspan can be five and a half to six and a half inches. It was a bit bigger than the one I had seen in the Essig Museum collection. The two large eyelike spots that gave the butterfly its name were even more convincing as they seemed to stare back at me, and I understood more deeply why the spots protect the butterfly from some predators. There are several species of moths with a similar kind of camouflage, like the ones seen in Bradovich's *Mothloric Muse*. The chrysalis of an owl butterfly looks like an autumn leaf folding in on itself. Another brilliant disguise.

The chrysalises in the sanctuary were as diverse as the butterflies. A few of these secretive chambers revealed an indistinct wing shape, but many completely concealed the contents and the messy work of metamorphosis. The transformation is hardly simple. The chrysalis is a miracle in itself, a sheltering shroud the caterpillar makes for itself as it transforms, its vessel for monumental change—changes impossible for me to truly grasp. The caterpillar digests itself until what is left is essentially a soup of vital cells from which emerge eyes, antennae, legs, and wings!

On my mother's refrigerator is this quote: "Just when the caterpillar thought her life was over, she became a butterfly." This metaphor has certainly been overused over time, but seeing the mystifying array of tiny butterfly containers made me conscious of the true miracle of this process. Inside a chrysalis, the caterpillar essentially disintegrates, becoming a fertile liquid in which float a few clusters of signal cells. These cells help reorganize the contents of the chamber into a new being, radically different from the one left behind. All of those cells reorganizing, all of the elements of one thing rearranged to be another. Like all of us earthlings made up of cells, of minerals, of elements—just in different patterns.

The work of the writer Cal Angus speaks so beautifully of earthly complexity in a passage from "The Moonsnail," a fictional account of Gertrude Stein's time at Woods Hole, Massachusetts, in which Stein examines and wants to deeply understand her discoveries at the beach: "She wanted to slow down and think more about keratin, limestone, calcium, break it all down piece by piece like a moonsnail drilling a hole in the shell of a smaller snail and flooding its caustic juices before slurping it all out. She would like to do that to the world around her. Liquefy until it bleeds together." As I read that passage, it struck me that this particular work of looking deeply, taking things down to the smallest particles of which everything really is made, might be a way

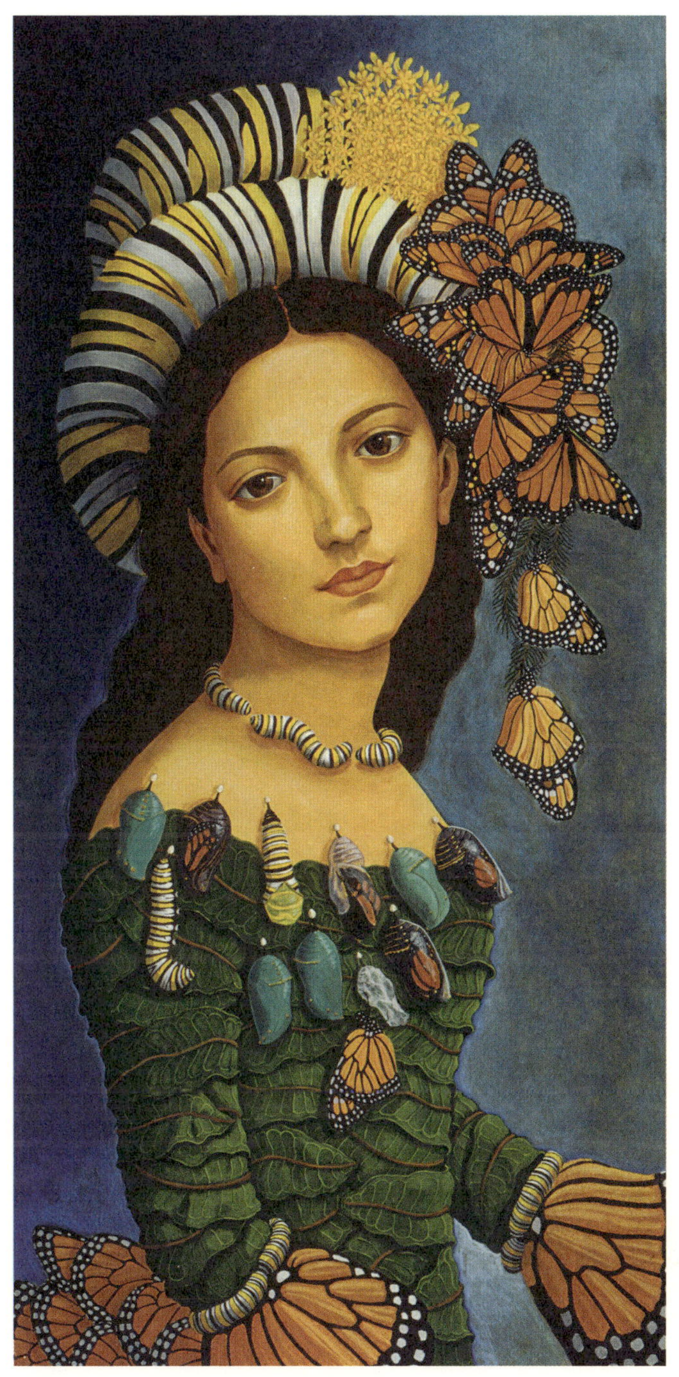

Lea Bradovich, *Metamorphosis Dress*, 2008.
Acrylic gouache on panel, 36 × 18 in.

to build empathy. Could this kind of close observation build connections between unlike things?

Bradovich's paintings also invite us to make connections, to put aside our fear of insects and bring them instead into intimacy. Virginia Woolf does something similar in her famous essay "The Death of a Moth," where she watches a moth as it struggles at the end of its life and then muses, "It was as if someone had taken a tiny bead of pure life and decking it as lightly as possible with down and feathers, had set it dancing and zig-zagging to show us the true nature of life." Bradovich invites us to see these miraculous shapeshifters and, perhaps, to transform ourselves.

In *Dragonfly Damselfly*, Bradovich depicts another human figure surrounded by insects. In this painting, a woman wears a garment of water. Water lilies float on the fabric of her dress. Dragonflies rest on the top of her head, her finger, her neck. A frog seems to be leaping out of her chest. She wears a watery camouflage, perhaps to hide her humanness, to appear closer to the nonhuman world, so that she can find connection with these insects. The dragonfly is also an insect that undergoes radical transformation, metamorphosing from an aquatic nymph into an aeronautical phenomenon.

Our human fascination with metamorphosis is ancient. Ovid begins his epic retelling of myth entitled *Metamorphosis* this way:

Bodies, I have in mind, and how they can
change to assume
new shapes—I ask help of the gods, who know
the trick:
inspire me now, change me, let me glimpse the
secret
and sing, better than I know how

Many of us long for change or want to cling to the possibility of transformation. Especially in this historical moment, in which we face a climate crisis and mass extinction, it seems we may all need these symbols of hope.

In Lea Bradovich's drawing *Heart of Gold*, a woman sits calmly though bees have taken over her body. The honeybees have made a comb of her heart; her lungs are a lattice of wax and honey. The bees do their work inside her, yet her eyes gaze directly at you and you see she is unafraid. Her arms are encased in bark as she becomes tree and bee home. She branches and blossoms, her heart now made of gold. Alchemy. Metamorphosis. Dreaming a future of interbeing, a seamlessness of earth and human. What would happen if we allowed ourselves to dissolve the human ego, letting go of the dream of the human as above and separate, and instead, humbled ourselves . . . opening and opening until at last we recognize the sacredness of each part of one vast and miraculous whole?

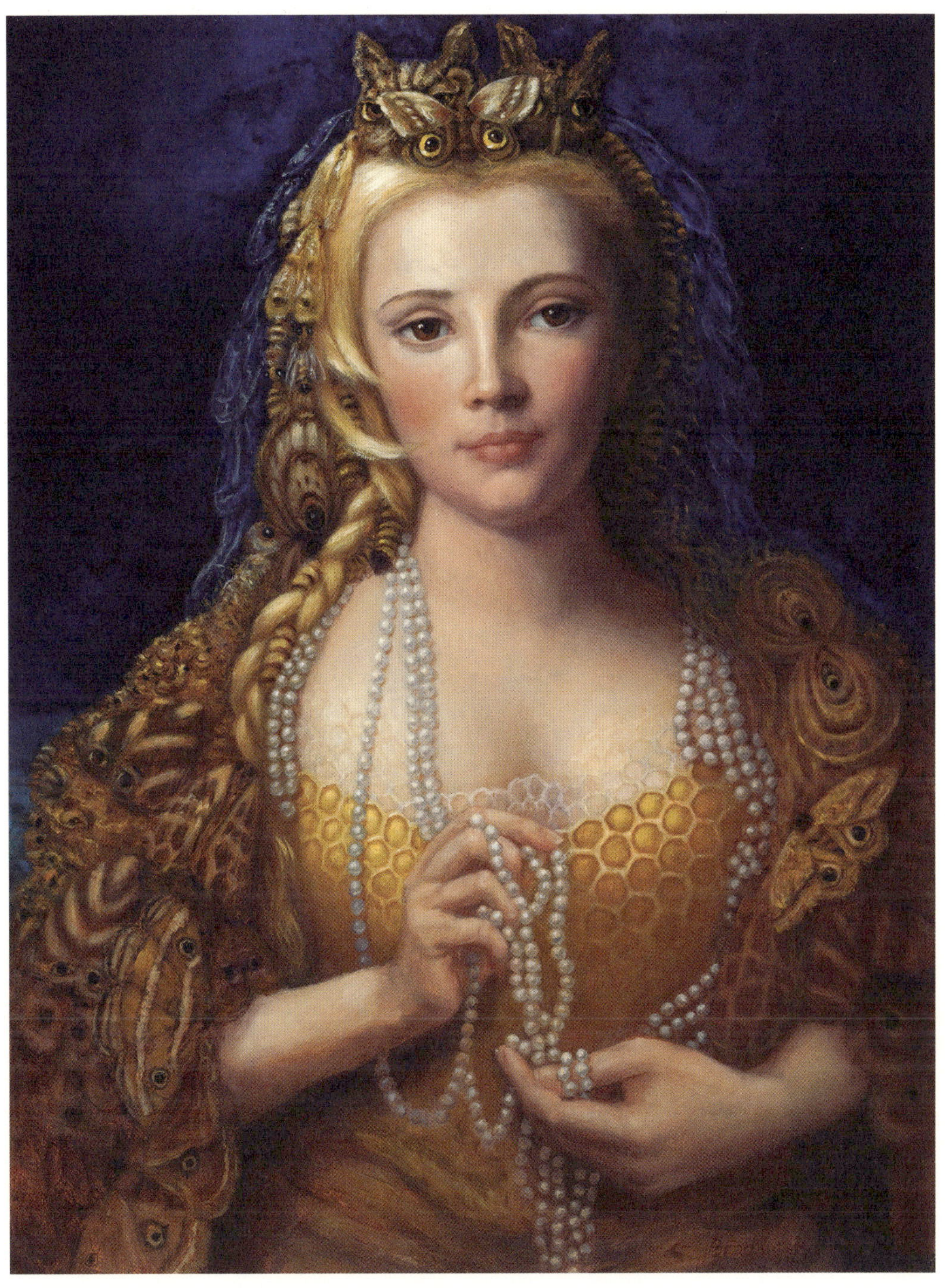

Lea Bradovich, *Mothloric Muse*, 2018. Oil on panel, 24 × 18 in. Photo: John Vokoun.

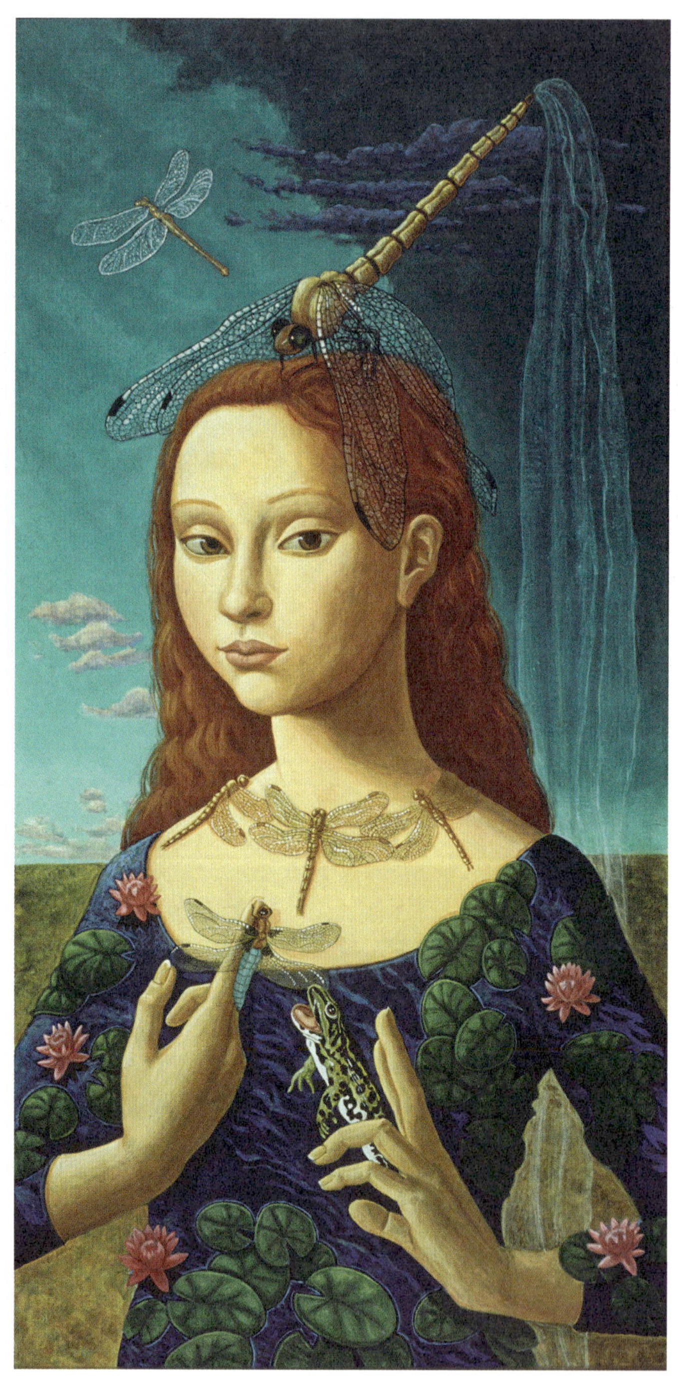

Lea Bradovich, *Dragonfly Damselfly*, 2008.
Acrylic gouache on panel, 24 × 12 in.

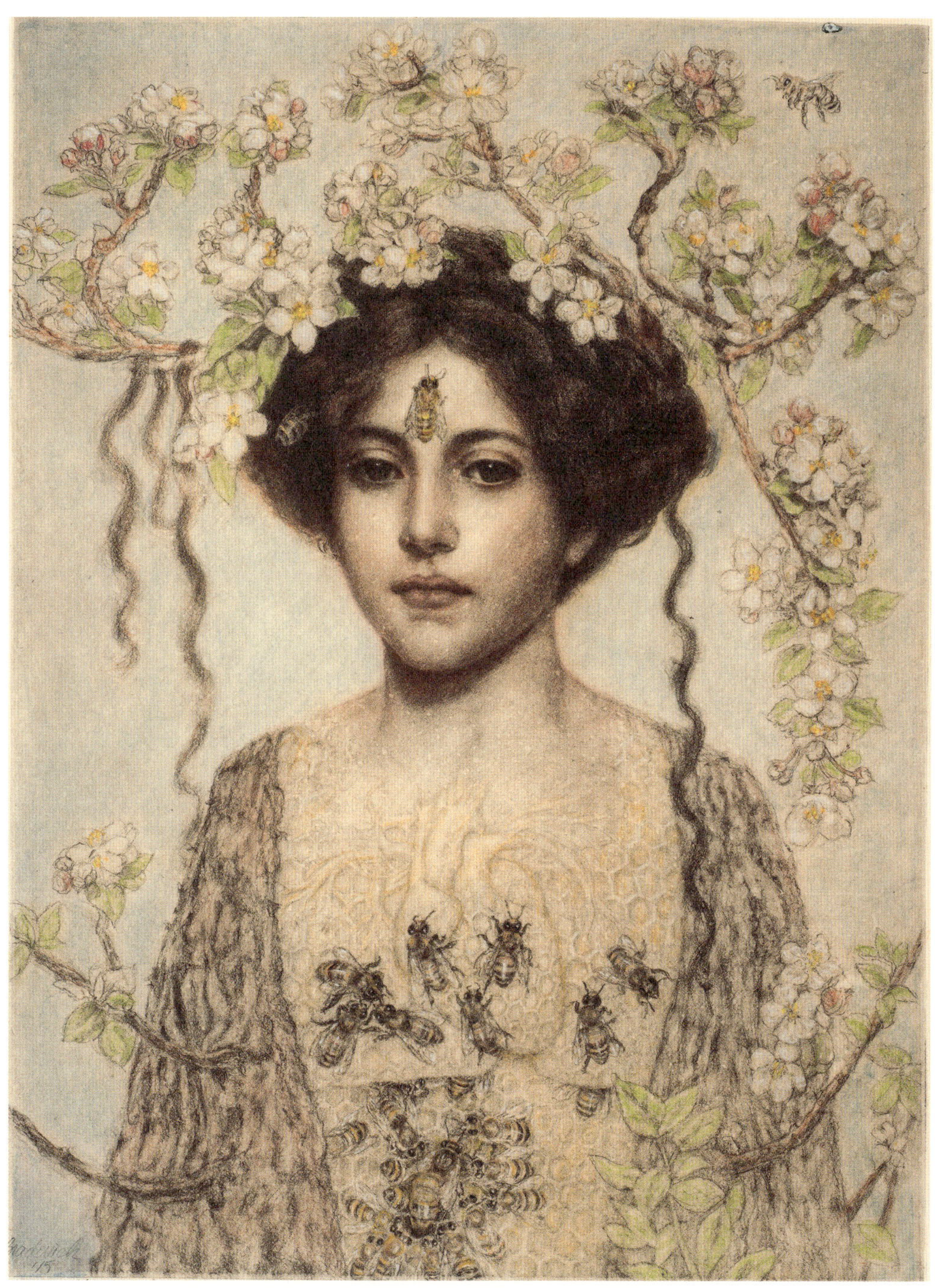

Lea Bradovich, *Heart of Gold*, 2020. Hand-colored photo etching, soy ink, and pastel, 16.5 × 12 in. Photo: John Vokoun.

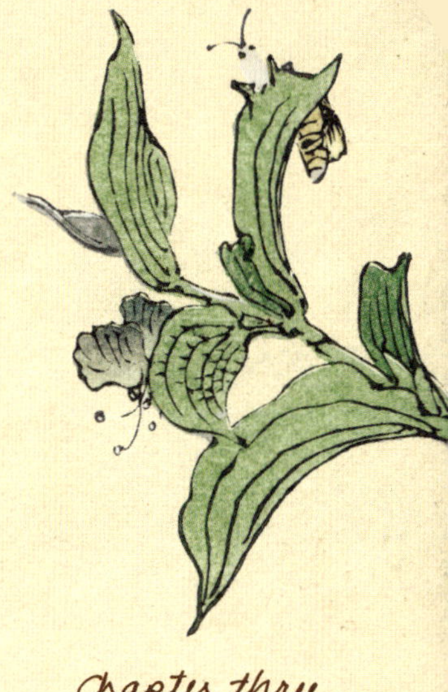

The Forest of Orchids

SOWING SEEDS OF RESILIENCE IN COLOMBIA

chapter three

> To him who keeps an Orchis' heart—
> The swamps are pink with June.
>
> —EMILY DICKINSON

Hidden in the verdant mountain forests of Colombia, among trees threaded with the flights of tanagers, iridescent hummingbirds, and a vast array of butterflies and bees, are some of the most seductive, aromatic flowers in the world: cattleya orchids. Cattleyas are known for their alluring scents and voluptuous, colorful blossoms and are sought by orchid lovers around the world. In 1936, Colombians proclaimed *Cattleya trianae* to be their national flower—an orchid named after the Colombian naturalist José Jerónimo Triana. The country is home to over four thousand types of orchids, which have flourished there for centuries, but today, in part because of their popularity and in part because of environmental destruction, at least 50 percent of these native orchids are endangered.

I confess I had some fears when I landed in the mountain city of Bogotá on a misty October night. It was the fall of 2019, and I had traveled to this Andean city to meet a family of self-proclaimed orchid addicts. My images of this country had been shaped by both fact and fiction. To many people living in the United States, in the past few decades, the mention of Colombia has conjured images of traffickers, guns, and suitcases full of cocaine under a thick canopy of trees—stories the media tends to highlight.

Although I had close Colombian friends, this was my first experience of this land. When I arrived, a cab took me through the busy streets of Bogotá to the neighborhood where I would be staying. The tips of the Andes—dark giants at the edges of the city—stood watch. This did not seem like a country that had suffered a violent fifty-year war and the complex problems that came with drug trafficking. In 2016, three years earlier, a peace accord had been signed, and the country was still trying to heal. Many Colombians, like the family I was headed to meet, were working to change their country's story. I knew those mountains held much more than coca plantations and guerrilla fighters. That night I dreamt not of violence and bloodshed but of forests, tangles of towering dark trees and flowering vines, singing with lyrical nonhuman voices.

The next morning, sunlight spilled over the mountains, illuminating the city. Shopkeepers and newspaper vendors readied for the day. Office workers hustled across streets or hailed cabs. Buses rumbled past. Flower stalls flaunting their colorful wares in tall silver buckets brightened the street corners.

Colombia is the second-largest flower exporter in the world. Often the roses we buy for our sweethearts on Valentine's Day in the United States have traveled from the greenhouses outside of Bogotá to our supermarkets. Flower exports bring approximately $1.5 billion to Colombia annually, and about 75 percent of those flowers are sold to the United States.

But the vast majority of these cultivated flowers are grown by a method that does not help preserve or restore Colombia's biodiversity. Like other industrial-scale, single-crop models, dependent on pesticides and fertilizers, these flower farms have devastated soil health, insect health, and ecosystem health. I had come to Colombia to learn about a different means of flower cultivation, one being carried out on a granular, species-specific scale on a rural mountainside. I had come to meet María Luisa Hincapié and her family and to learn about their project: the Forest of Orchids.

Colombia holds the distinction of being one of the most biodiverse countries in the world, with five biogeographical regions, nearly 2,000 types of birds, 20 percent of the world's butterflies, and 734 species of jewellike frogs. Many of these species are endemic—belonging to unique, highly specialized ecosystems—but deforestation, mining, the disappearance of pollinators, and illegal trafficking are destroying these fecund spaces.

María Luisa, her husband, Luis Carlos, and their daughters, Mar and Sol, had embarked on a journey to protect one small corner of this fragile world by restoring a mountain landscape that supports native orchids and their pollinators. María was familiar with my first book as well as with some of my other writing about the relationships between insects and plants, and she had invited me to visit so that she could show me her orchids and her land.

Orchids have captured the attention of humans for centuries. The extraordinary plants are known for their intricate blossoms, arresting scents, and bright colors. Many of those found in the mountains of Colombia are epiphytic—which means they attach themselves to trees and seem to miraculously live on air—but the orchid family is so broad and diverse that you can find orchids in deserts, on grasslands, in the chilly climates of Canada and Siberia, and even living underground. Their flowers have invited many interpretations. The most familiar blossoms—with their colorful petals and distinctive lip opening to a deep cup—resemble the female reproductive anatomy. But the word orchid or "orchis," according to the *Oxford English Dictionary*, comes from "modern Latin, based on Greek *orkhis*, literally 'testicle' (with reference to the shape of its tuber)." In the 1842 edition of *The Lady's Book of Flowers and Poetry: To Which Are Added, a Botanical Introduction, a Complete Floral Dictionary; and a Chapter on Plants in Rooms*, Lucy Hooper explains that "the Greeks named this plant Orchis, from the form of the roots in many of the species. . . . In addition to the Greek name, the Latins often call it Satyrion, because the early Romans believed it to be the food of the Satyrs, and that it excited them to the excesses which in fabulous history are ascribed to them." Even today, orchids are thought to be powerful aphrodisiacs.

The flowers are more than our erotic interpretations of them, however. Vanilla orchids give us an unmistakable vanilla flavor integral to so many of our cookies, cakes, and pastries. A research center at the University of Washington in Seattle claimed that some orchids "are made into a remedy for sick elephants in Malaysia and a kind of ice cream in Turkey . . . said to prevent cholera, heal the spleen, and ease childbirth." According to *China Daily*, "Ancient Chinese called eternal friendship '*lan jiao*,' or companionship as nobel [*sic*] as the orchid, and beautiful prose and poems '*lan zhang*,' or words as graceful as the orchid." Some of these flowering plants can live up to one hundred years, and their cultivation has been practiced for centuries. The Japanese poet Izumi Shikibu, from the Heian court of ancient Japan, wrote:

> As I dig for wild orchids
> in the autumn fields,
> it is the deeply-bedded root

that I desire,
not the flower.

In Victorian England, an obsession with orchids, especially rare ones from colonized lands, led to an enterprise of feverish and illegal extraction—one of many imperialist ventures of the time. In *Orchid Fever: A Horticultural Tale of Love, Lust and Lunacy*, author Eric Hansen recounts tales of incredibly intrepid and obsessive orchid collectors, tales that end in fights and sometimes death. Even today, precious or rare orchids fetch fantastic prices. A single Shenzhen Nongke orchid, developed in China, can cost an orchid collector $202,000. High prices make the illegal removal of native orchids tempting despite the danger of obtaining them. The value of illegal native orchid trafficking globally is estimated to be $6 billion annually.

When I traveled to meet María Luisa, I admit, I knew little about orchids other than that they were exquisite flowering plants that required special pollinators. Each variety attracted a different bee or wasp or moth or butterfly to pollinate it, and I wanted to learn more about this.

María Luisa spent countless hours of her childhood outside. The natural world was her playground. She loves telling the story of when she and her cousins befriended an alligator hatchling, which she kept under her bed at night until they realized that wild animals should be allowed to live in their own habitats, not in the possession of a human being. But it continued to visit them, sometimes knocking its head against her door to get her attention, even after she'd returned it to its home. This experience has stayed with her and informs her respect for the land and its creatures to this day.

I never had a pet reptile, but when I was eight, I fell in love with honeybees. My father took me to a honey harvest at a friend's small apiary. The sweet perfume of honey and wax filled the air as bees floated around the gentle beekeeper. He took only the excess honey, leaving enough for the bees to survive the winter. The relationship I saw between bees and beekeeper seemed mysterious to me. I dreamt of one day keeping bees myself, and when a young woman I got my first hive. I never had any interest in selling honey or renting out the bees for pollination services. I simply loved watching them nuzzling flowers, dancing their maps to one another in the hive, stroking each other with their antennae. I admired the fact that their way of life was truly one of reciprocity. They took nectar from the flowers but pollinated them in return. They lived so gently in their communities.

Honeybees are the gateway bug, my mother often says. You can capture someone's attention with the honeybee and then open them to the miracle of all the other insects. My interest in honeybees led me to research native bees, dragonflies, moths, crickets, and eventually the astonishing array of creatures that pollinate orchids. The insect world and all of its connections to landscape became my obsession. But as my awareness and awe expanded, I simultaneously grew more alarmed. Many pollinators and other insects were seeing rapid decline.

As I mentioned, I learned about a study published in *PLOS One* by Caspar Hallmann and his colleagues reported a loss of 75 percent of flying insects across Germany over a twenty-seven-year period. In 2019, another study, published in *Biological Conservation* and incorporating the historical data from seventy-three sources worldwide, concluded that 40 percent of insect populations were showing rapid decline and might be facing extinction. Among the insects mentioned were butterflies, bees, dung beetles, and dragonflies.

A recent *PNAS* report called "Insect Decline in the Anthropocene: Death by a Thousand Cuts," which references more than eighty other papers from around the globe, notes that insects are crucial for more than just pollination. They are also important as food for other species in complex food webs, as consumers of smaller insects that can threaten human quality of life, and as agents in "the macro decomposition of leaves and wood and removal of dung and carrion, which contribute to nutrient cycling, soil formation, and water purification." "Clearly, severe insect declines can potentially have global ecological and economic consequences," the authors conclude. The report also states that "the principal stressors—land-use change (especially deforestation), climate change, agriculture, introduced species, nitrification, and pollution—underlying insect declines are those also affecting other organisms. Locally and regionally, insects are challenged by additional stressors, such as insecticides, herbicides, urbanization, and light pollution."

Reports like these continue to prove that many insects are in trouble. What does this mean for our world? In *On the Origin of Species*, Charles Darwin wrote about the abundance of red clover in a certain region of England as an example of the interconnected relationships among species. The clover was pollinated solely by bumblebees, and it was abundant there because the bees were abundant. The bees, in turn, were dependent on the presence of cats, who devoured the mice who disrupted the nests of bees.

To get to the Forest of Orchids nature reserve and lodge, I drove from the sprawling city of Bogotá out to a small town called Tenjo. As I left the urban area, valleys opened out into pastures and fields. I eventually arrived in a town nestled against one of the surrounding mountains. I navigated through the narrow streets of Tenjo to the edge of a dramatic rocky incline. Though the GPS indicated I had arrived, there was nothing that looked like a lodge in sight. There was, however, an incredibly steep, muddy road that disappeared around a bend. I decided to get out and walk. A few small dogs swarmed around me, barking loudly, as I came to a tin-roofed house perched on the side of the mountain. A young girl hanging shirts on a clothesline in an open-air kitchen paused to call them off. I asked about the reserve, and she pointed straight up. A gate "a la derecha . . . no lejos," she said.

It was a steep incline. As the road leveled out slightly, a large German shepherd rushed toward me, scolding me with a ferocious bark for approaching the gate. A second one from inside the enclosure began barking, too. From somewhere inside the gate came a melodic female voice calling to the dogs in Spanish. Then I heard, "Adelante, come in!" I had arrived at the reserve. A wide, radiant smile flashed under María Luisa's high cheekbones as she opened the gate for me. Her dark eyes appeared both fierce and full of love.

When María Luisa's family bought the land in 2001, it had been scraped clean of any vegetation. The soil was dry and impoverished as a result of overgrazing. Since then, her family has done the work of restoring the native vegetation of the mountainside by planting a native orchid reserve—part of their shared vision to change the story of Colombia from one of violence and destruction to one of restoration and healing. With orchids.

Beyond her verdant hillside, María Luisa's country was recovering from more than overgrazing and the illegal native orchid trade. Colombians were trying to heal from the long civil war that began in 1964 and continued until 2016, when, finally, a tentative peace accord was signed between the FARC (Revolutionary Armed Forces of Colombia) and the government. The conflict involved asymmetric warfare in which the government fought not just left-wing guerrilla groups, such as the FARC, but also right-wing paramilitary groups in addition to drug cartels and other organized crime. Throughout those decades, several violent factions struggled for power—a struggle that was intensified by the booming demand for cocaine within the United States—and civilian life was shrouded in a veil of fear. Estimates suggest that over 220,000 people were killed during this time.

Civilians were caught in crossfire that rang out daily in urban neighborhoods; bombs were placed in cafes and nightclubs; public figures were kidnapped. And rural towns were not immune from the violence. Frequently one armed group or another would overtake a small farming community and force residents to support a specific group or plant coca for the drug trade. Although there are signs of change, the fragility of life is a fact many Colombians seem to know deeply and intimately.

The light was fading fast at the Forest of Orchids reserve, so María Luisa and her daughter, Mar, quickly ushered me out to see her orchids. The tinny sound of a loudspeaker from a small truck selling something in the town below was nearly drowned out by crickets and the calls of crepuscular birds. The dogs stayed nearby, as if to remind me not to make any wrong moves. As she led me farther up the steep grass-covered paths, María Luisa explained that, though she and Luis Carlos had been spared the violence of the civil war on their land, the mountainside had experienced a different type of violence. When they purchased this property, she reminded me, it had been completely denuded. Many areas in Colombia had been stripped for mining or grazing or industrial agriculture. Environmental violence also plagued this country.

María paused in the late light and pointed to specific trees and bushes. "We planted all of this. There was nothing here but naked soil and rock." I glanced up at the hillside, which was now covered with young trees, shrubs, and orchids.

She stopped in front of a wall covered in plants with leaves of many shapes—long and thin, or short and thick—and no two had the same flowers. Each plant's blossoms were unique. "Which are orchids?" she asked. I suspected the answer. "Todos," she said. *All of them*. She told me that there were approximately thirty-seven thousand species of orchids in the world.

María and her family planted the Forest of Orchids not only to renew the land where we were standing but also to save native Colombian orchids, many of which are in danger of extinction. The orchids of Colombia face a number of threats, one of the greatest being illegal orchid trafficking. One reporter from Bogotá quoted the police as saying that in 2017 over two thousand species of orchids had been recovered from smuggling operations. As habitat is destroyed, pollinators become scarce. As the diversity of insects decreases, so, too, does the diversity of orchids.

The Forest of Orchids has four thousand species of orchids growing in its greenhouses and naturalized on the land. When researchers at the National

Museum of Natural History in Paris decided that they wanted to follow in the footsteps of nineteenth-century naturalist Alexander von Humboldt, a pioneer in ecological thinking, to discover whether the Colombian orchids he wrote about were still living in the same regions, the head of the project contacted María Luisa, Mar, and Luis Carlos to lead the expedition. The expedition took them deep into other parts of Colombia, where they found some of the very flowers described by Humboldt. The reputation of María Luisa and her family as orchid experts and conservationists continues to grow internationally.

But their work faces serious obstacles. Orchid thieves are not all they have to worry about. In a recent article in *Edge Effects*, the environmental scholar Rob Nixon illuminated the great danger of being an environmental defender: "In the 245 weeks since the 2015 Paris Climate Agreement was signed, an average of four environmental defenders have been murdered each week." Colombia, specifically, is listed as among the most dangerous places in the world for environmental activists. In 2019, sixty-four Colombian land defenders were killed because of their work.

We walked farther up the hillside, where María Luisa reached into a tree and stroked the long leaf strands of an orchid, explaining how she attached them to host trees. The orchid did not need soil but could obtain the nutrients it required from the air around it. At first there were few insects at all, she said, but as they planted new plants and let some fallen pine trees decay, beetles began to return and restore the soil. Fungi began to do their magical work of nutritional networking as the mycelium mingled with the many roots. The renewed fertility of the soil allowed windblown seeds to find purchase, and so the natural cycles of the mountain began to return.

In Andrea Wulf's biography of Humboldt, she summed up his observations on biodiversity with this: "Nature's balance was created by diversity which might in turn be taken as a blueprint for political and moral truth. Everything, from the most unassuming moss or insect to elephants or towering oak trees, had its role, and together they made the whole. Humankind was just one small part."

Over the course of nearly two decades, the slopes of María's mountain transformed. Butterflies had returned. Bees had returned. Now, she said, birds, frogs, and mammals were returning to the mountain as well. She pointed to some low shrubs. These were native volunteers, she explained, smiling. "And now people come to see them." I paused to take in the life pulsing around us—María Luisa and Mar, mother and daughter, both lithe and almost otherworldly in this light, standing on this verdant land that had once been dead. They themselves seemed orchid-like, able to pull sustenance from the air around them.

The mountain was falling into shadow as we wound back down the incline toward the nursery: essentially a roof with fabric walls and openings for insects to come and go. This building was home to all of the orchids the family had collected and was propagating in order to repopulate the mountainside. Inside, María Luisa leaned in close to the plants, caressing them and speaking quietly about the womb of the orchid, a small pouch that would thicken after pollination, and the most important part of the plant, in her opinion.

"Orchids coevolved with insects, so they each have special traits," Mar explained from the doorway, and she told a story about an orchid that lived by the sea. It had evolved so closely with its specific environment that it had developed a scent of rotting fish, attracting the flies inhabiting that particular shore. Joe Meisel's book *Orchids of Tropical America: An Introduction and Guide* illuminates dozens of strange orchid/pollinator relationships. There are so many different genera in the orchid family. *Elleanthus*, for example, have long, tubular deep-red or purple blossoms, which are a perfect vessel for the long beaks of hummingbirds, while *Brassia* have long, thin sepals that look very much like arachnid legs. The flowers are pollinated by large wasps called tarantula hawks, which hunt large spiders, stun them with their stingers, and drag them to their burrows, where they lay an egg on their bodies. The *Brassia* mimics the spider body, tricking the wasp into trying to sting the flower. During this action the wasp picks up pollen and carries it to the next flower.

María Luisa is not only amazed by their ingenious evolutionary distinctions but is also convinced that plants are intelligent in ways that we do not understand. They "see and know different things," she said. "They know much more than we do in some ways." I asked her if this meant that she thought plants had consciousness. She gazed at me, measuring, and then nodded. "Yes, I think so. I talk to them, and they exude chemicals that can change us," she said. She smiled like a person with a secret. I thought of the purported magic of cattleya. María Luisa walked to a smaller plant. "Once a woman came to visit us after her husband had died a long and terrible death, and she spent time touching this particular orchid. The next day the plant withered and nearly died. But it came back." María Luisa believed that it had absorbed her despair.

The idea that plants are sensing beings is not new. Many Indigenous Colombians, like the Wiwa of the Sierra Madres, believe that rivers and mountains are alive, that the animals and plants are family. These beliefs have existed for thousands of years.

Researchers like Suzanne Simard and Merlin Sheldrake have done extensive research on the mycorrhizal networks that connect plants. Hub trees can steer nutrients to their offspring. Some trees emit warning pheromones to nearby trees when they are attacked by a predator insect so that the neighboring trees can begin to emit an armor of tannins to repel the same insect. Trees even have electrical impulses they release when wounded. But are plants able to sense human energy? Can they help us heal? Inspire us to heal the earth?

I believe that bees, too, are sensing beings and that they have moods. They are so perceptive and sensitive with one another and seem to be able to read me as well. I've often sat for hours watching honeybees gently pollinate flowers and then return to their hive and dance maps, sharing the location of newly discovered nectar. Often they stop to caress each other's faces or try to help another bee who is struggling, exhibiting what looks like love. The bees remind me that communities can live harmoniously with one another. I got into the habit of talking quietly to them at the hive and soon was whispering to native bees when I walked in the fields. But I also tried to listen.

Inside the Hincapiés' home, the three of us were joined by Luis Carlos. "It all started because of a birthday present of an orchid to my father," Mar said. "And then the addiction began." She was smiling. "Here is a special orchid," Mar continued. "This one is named after my mother! Together we discovered a new species. This one is called *Maxillaria maria-luisae*."

Luis Carlos's passion for orchids drove him to research them. He loved the science of orchids, which he had shared with his daughter. Mar showed me photos of the many types of birds, orchids, and bees they had seen on the land. Luis opened a book and showed me a photo of an orchid thought to be the last of its kind, an orchid they had on the property: the *Ida lionetti*. They had found one—just one—growing on the mountain, when the vegetation was still sparse and recovering. Luis had identified it by its characteristic peach-colored petals and waxy leaves. Orchids can live a long time, scores of years, and so they waited. Finally, after the mountain had come to life again, the right pollinator arrived and "impregnated" the plant. Once it was pollinated, they took the seeds and propagated them, saving this orchid from extinction.

Mar called me over to a table where she had placed a small collection of glass bottles, vials, powders, and plants—a scene from a mad scientist's laboratory. She explained the incredible patience it takes to grow orchids. Each seed is minuscule, like dust, capable of being carried by wind. Thousands of them are harvested from the "womb" of the flower and placed in one of these bottles, along with a very special type of algae that allows them to germinate. The seedlings grow and then harden. This process can take up to two years! Eventually they are transferred to a new habitat, but that habitat has to have the perfect microclimate and the perfect pollination assistants.

Luis Carlos offered to demonstrate the pollination of an orchid that was blooming on their dinner table. "Sí, por favor," I said, and everyone grew very quiet, solemn, and still. I realized that this was a sacred action. With one hand he lifted a white blossom and with the other he held a long, thin pipette. As he slowly moved the pipette into the open blossom, he whispered about the sexual organs of the flower. He pointed out a cap on the tip of a long shaft under which the pollen was hidden. Only by releasing this pollen could it reach the female parts of the flower at the base of the shaft. If an insect was too big or too small, it would not trigger this process: "The orchid would not get pregnant." The room was quiet. He gently touched the flower and it sprang open.

"Incredible," he uttered, staring at the flower like a man who had just seen an orchid for the first time. I had seen bumblebees resting in flowers in what looked like ecstasy. I thought of a poem by R. Snow in Hooper's *Lady's Book of Flowers*, which describes the orchid bee:

> Admire, as close the insect lies,
> Its thin-wrought plume and honey'd thighs:
> Whilst on this flow'ret's velvet breast,
> It seems as though 'twere lull'd to rest

Indeed, Luis Carlos seemed lulled to rest. "He is addicted," his daughter whispered.

Caring for orchids and repopulating the forests has been a huge effort, but passion made it possible. The Hincapiés explained that it is part of a larger vision. When I asked Mar, the youngest among us, why she did this work, she said she felt called to protect the mountain. And, more broadly, to change the story of Colombia from one about war and drug lords like Pablo Escobar to one that shows another side of Colombia: the nature, the literature, the art. "Gabriel García Márquez wrote magical realism because it is magical here," she asserted. Light seemed to be emanating from her body. "It takes a lot of love for the healing," she said. "A lot of love."

Late that night, as I descended the mountain, a light rain fell on the Forest of Orchids. As I made my way back to Bogotá, I thought about the noble endeavor of restoration in a country plagued by violence. Decades of unchecked terror could leave one immobilized by fear, disconnected. But María Luisa's family had chosen to restore a mountain, which they believed could help change the story of Colombia and make a different future possible. María Luisa told me, "Nature teaches us about resilience, diversity, hope, second chances, and healing. . . . Starting with eroded land, transform[ing it] into a small forest full of life, and being able to witness the return of many native species—as well as being able to live in harmony with other species—is something unique and wonderful."

But there is no guarantee that the transformation they have helped bring about will endure. In February 2020, a wildfire burned through a part of the Forest of Orchids reserve and destroyed parts of the land that María Luisa and her family had worked hard to regenerate for eighteen years. When María Luisa told me about it, she said that they were already in the process of replanting. "Sometimes, when we feel like giving up," she said, "an orchid that blooms for the first time or a hummingbird that flies around us gives us again hope."

María Luisa's time on the mountain has taught her that we all have an important part to play in an interconnected world, and she is deeply in love with that world. She knows that successful ecosystems depend upon biodiversity and, while humans are in no way at the top of any hierarchy, we can place our hands in the soil to help; we can become one strand in the great web; we can help ensure that the next orchid blooms.

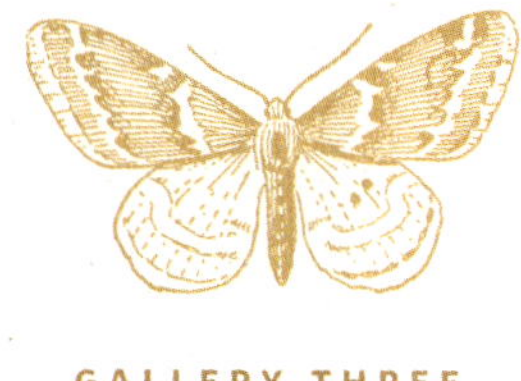

Susan Carlson

Quilting a New Reality

Quilting is a way to transform what has been left behind. Quilters take scraps of fabric and make something new, something beautiful. Susan Carlson's quilts stand out as especially spellbinding, not only because of their electric combinations of color and print but also because they contain recognizable images. Among her images made with a mosaic of fabrics are animals such as a polka-dot dodo, a twenty-foot saltwater crocodile, a sea turtle swimming in a sea of aquamarine, a red-and-pink rhinoceros, and a host of delightful insects. When I discovered them, I was swooning.

Carlson studied at the Maryland Institute College of Art, where she earned a degree in illustration and experimented with collage. Her mother is a seamstress, and Carlson believes that her love of fabric, the tactile aspects as well as the excitement of placing colors side by side, came from her childhood. While traditional quilters sew bits of fabric together, Carlson adopted a practice of gluing them together before quilting the entire piece, which gained her an international reputation in fabric collage art quilts.

When she was a little girl, her favorite museum was the Smithsonian National Museum of Natural History, where she encountered insect collections and was thrilled "to see all the groupings of specimens, neatly arranged and labeled in their glass-topped frames. The colors and variety of shapes and sizes of these creatures—their otherworldliness." Later, when making her insect quilts, she began "to reimagine those insect specimens" in fabric. "I even 'boxed them in' with borders of background fabrics," Carlson adds, "as opposed to placing them in more natural settings."

Carlson's insects grew out of her admiration for their symmetry and beauty but also out of the myths associated with them. One of the pieces I feature here was inspired by Carlson's knowledge of African mythology—specifically, a creation story involving a dung beetle.

When I was a little girl living among painters, potters, and sculptors, my mother taught me to sew. I knew how to sew with a machine by the time I was seven. We made nightgowns, skirts, stuffed animals,

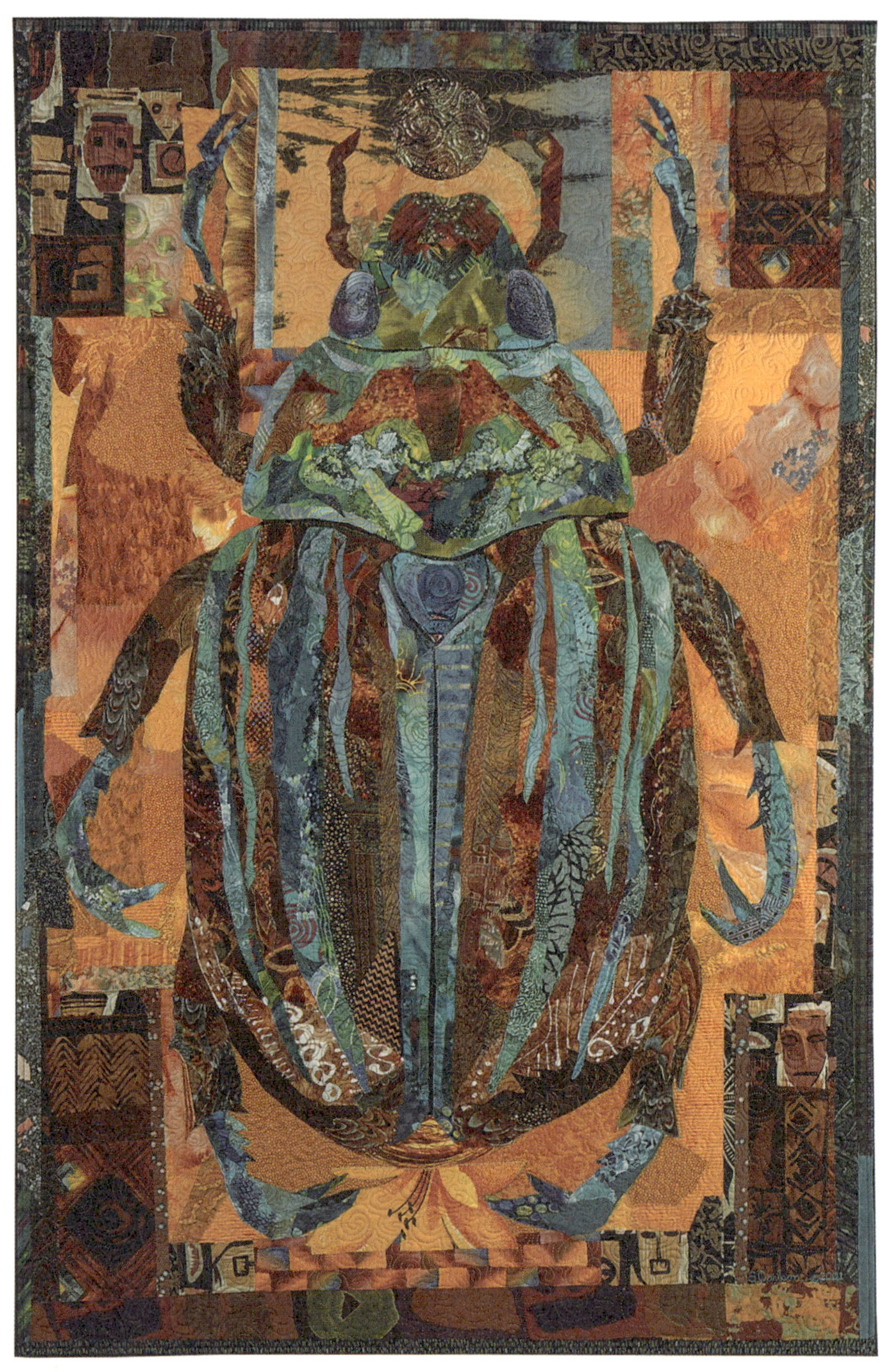

Susan Carlson, *Rolling Up the World*, 2001. Fabric collage, 49 × 33 in.
Private collection. Photo: Andrew Edgar Photography.

dolls, and shawls for my dog. When I was a teenager, I fell in love with the quilts made by the quilters of Gee's Bend, Alabama, a group of African American women whose talents had been passed down since the days of slavery. Traditionally, quilters picked up any scraps of worn fabric they found and turned them into blankets that could keep their families warm. They sang hymns in gorgeous harmony as they quilted together. Their resilience, their sense of community, their aesthetic, and their moxie inspired me. I decided to try quilting. This was a revelation to me: suddenly, I could bring together scraps of fabrics of different patterns and textures to make something of my own design. It was like painting with fabric, though my designs were always abstract and never anywhere near the sophistication of Carlson's work or that of the quilters of Gee's Bend. It did give me a deep appreciation of the craft.

But quilting gave me something more. On a subconscious level, it made me realize that diversity of color, texture, shape, and pattern, the juxtapositions of different components, made for a much more interesting final quilt—just as a diverse society or a diverse ecosystem is richer and healthier. Having moved a lot as a child, my own life was a kind of pile of beautiful scraps—memories of people and landscapes and experiences—that came together into a unified whole. Writing poetry, too, was not much different from this process of pulling fragments together, although I was using words. This book itself is a quilt of sorts. As I think about what our world needs right now as we confront so many urgent challenges on our planet, including failing ecosystems and fading biodiversity, it is not a singular, monochromatic idea. Rather, it is a mosaic of ideas and strategies and passions quilted together to make a more ethical and beautiful new. Mosaic thinking has become the way I survive.

That insects inspired Susan Carlson aesthetically is no surprise. Beatles and butterflies have some of the most astonishing color and design elements of any creatures on earth, rivaling the most brightly colored parrots. We tend to miss their beauty because insects are so small. Carlson's large quilts make us notice.

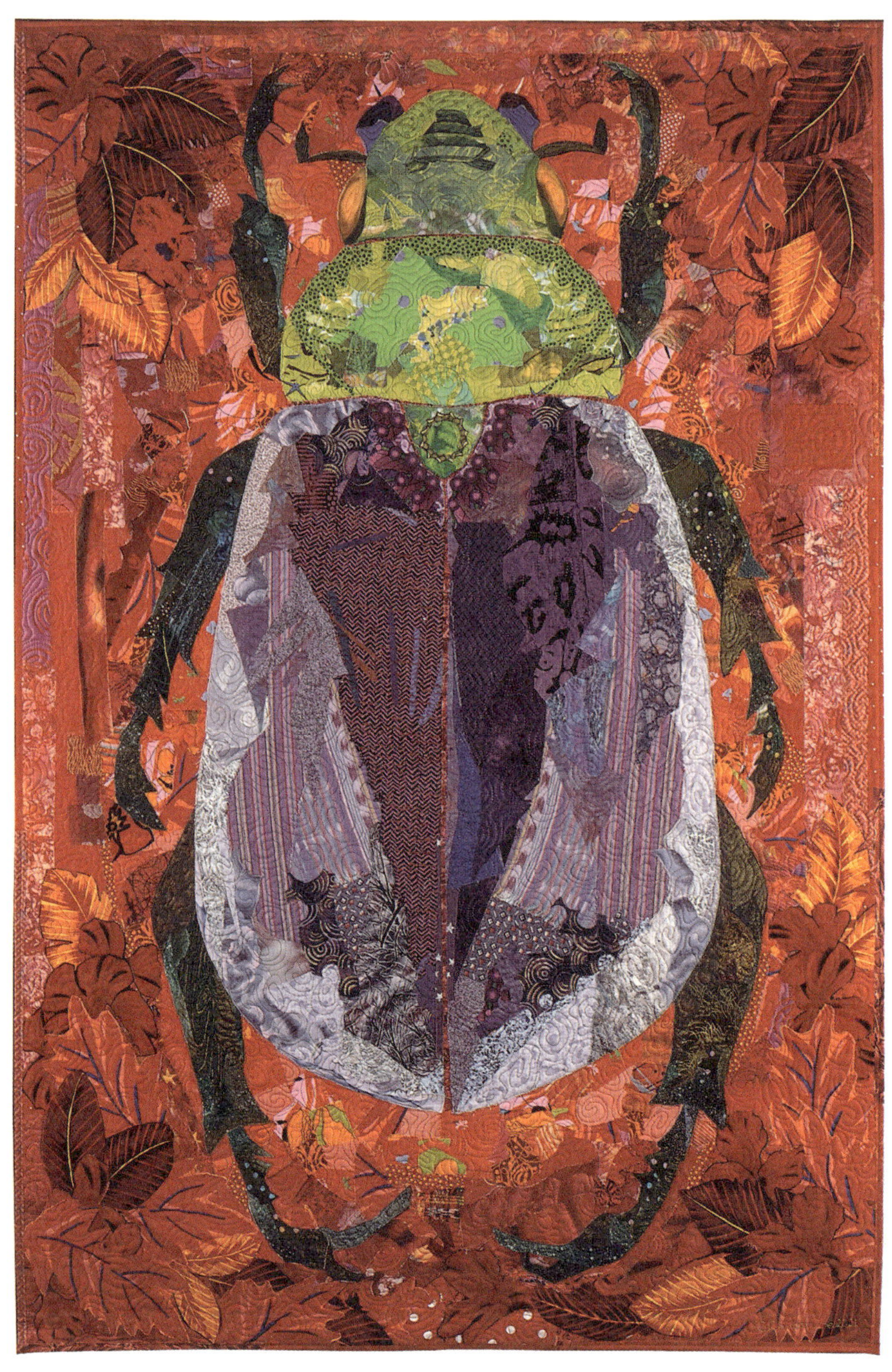

Susan Carlson, *Life in the Leaves*, 2001. Fabric collage, 49 × 33 in.
Collection of Pamela Holst. Photo: Andrew Edgar Photography.

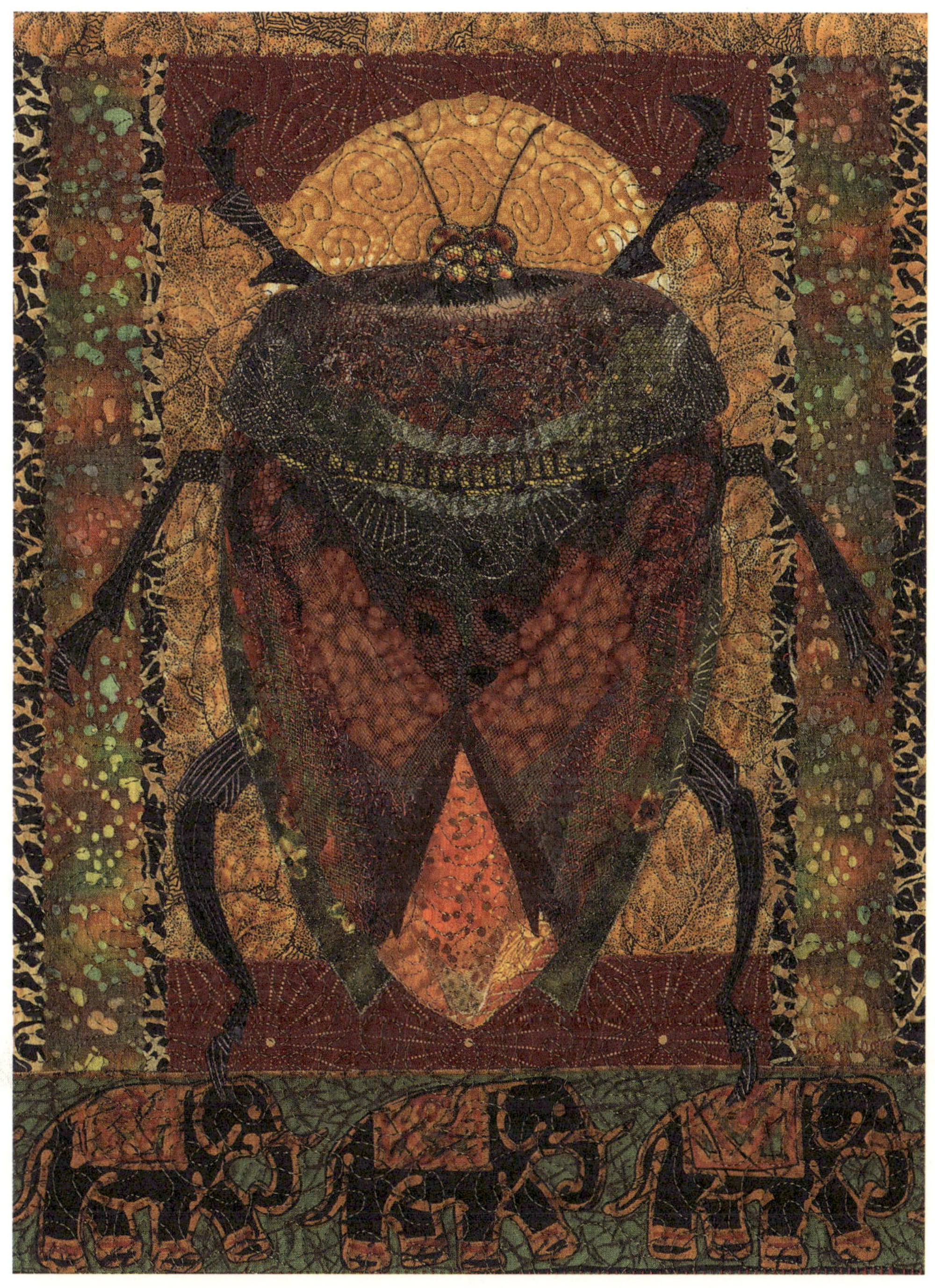

Susan Carlson, *Gargantua*, 2005. Fabric collage, 16 × 12 in. Private collection. Photo: Dennis Griggs of Tannery Hill Studios Inc.

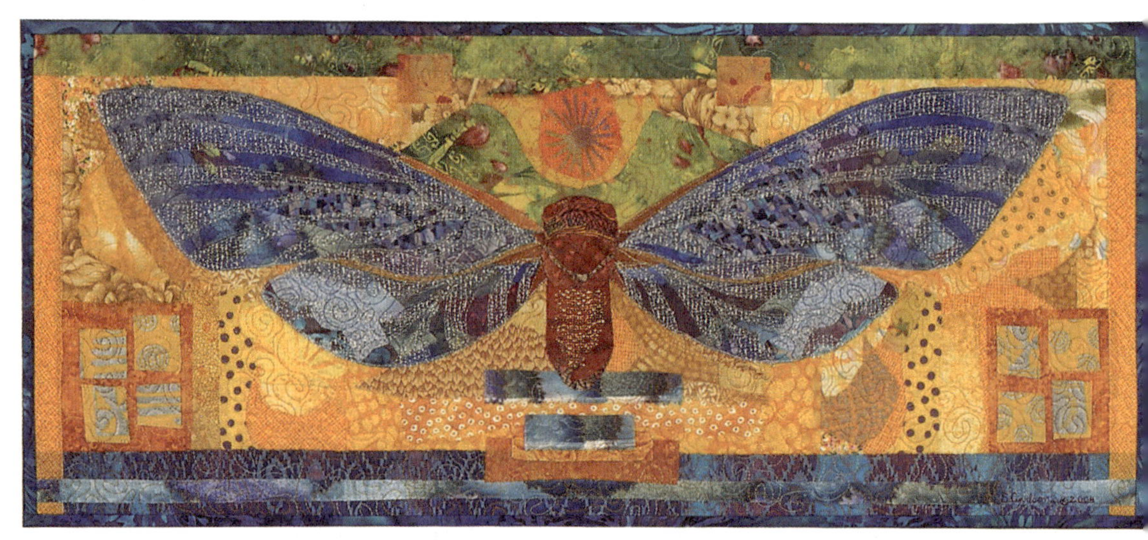

Susan Carlson, *Cicada Sunrise*, 2005. Fabric collage, 12 × 28 in. Collection of Leslie Smith. Photo: Dennis Griggs of Tannery Hill Studios Inc.

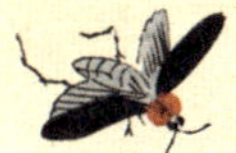

For the Love of Frogs

My discovery of Colombia's ecosystem restoration and conservation continued the day after I left María Luisa Hincapié and the Forest of Orchids. Next I would meet Ivan Lozano, a brave scientist on a mission to stop the illegal exotic pet trade—specifically, the trade in poison dart frogs. Colombia has not only good coffee and beautiful orchids but also the most beautiful species of frogs in the world. These frogs are dependent on the wealth of insects available in the rainforest, which is quickly disappearing. But even in the places where forests are intact, the frogs are in danger.

Lozano's red pickup sped through the tangled dark green forests of the Andes as we descended from Bogotá to a slightly warmer elevation where poison dart frogs thrive. Lozano, a zoologist, internationally renowned frog expert, and pioneer in innovative conservation strategies, was driving. He has an athletic build, a chiseled jaw, and black hair. He wore a black leather jacket and radiated positive energy. When I first met him, I quickly noticed the merriment in his smiling eyes and caught a glimpse of his sharp wit, which, I would learn, was key to his resilience. Lozano is also an accomplished race car driver, so we were making good time. The conservation projects he worked on here in Colombia could be incredibly stressful. Racing was his therapy, he said, grinning. His deft maneuvers around the switchbacks had me clutching my seat, but I seemed to be in very good hands as we headed to his frog breeding center, Tesoros de Colombia—The Treasures of Colombia.

Colombian dart frogs are like living jewels. Their small bodies of bright lemony yellow, tomato red, royal blue, or chartreuse with dark spots are arrestingly beautiful. They are also deadly. A drop of the venom from the skin of some poison dart frogs is enough to kill a human being. The venom from others is used in shamanic ceremonies to induce hallucinogenic journeys. Some people willingly pay hundreds of dollars for these tiny animals.

Many frogs—and amphibians more broadly—are disappearing with astonishing speed and facing extinction. Recent assessments find that close to half of all amphibians are at risk of extinction, and 2.5 percent are recently extinct. The alarming decrease in their numbers is due to a wide range of things, including habitat reduction, a depletion of insect food sources, a deadly fungus called chytrid, climate change, and waters poisoned with pesticides. In Colombia, which is home to the second-largest number of frog species in the world, the frogs face another danger: the illegal wildlife trade.

According to the US Fish and Wildlife Department, the illegal wildlife trade is a multi-billion-dollar enterprise. Colombian poison dart frogs are highly valued species, and their numbers are dwindling in the wild.

Lozano has dedicated his life to changing this trend by breeding endangered poison dart frogs both to restore the native populations and to undermine the illegal market for these frogs by selling them legally to pet stores. Illegal wildlife trafficking is his nemesis; battling it has kept him going despite the dangers he faces. Global wildlife trafficking can lead to violence, the spread of disease, and extinctions. Governments struggle to curtail the sophisticated crime networks. Lozano was fighting a huge problem.

After a couple of hours on the main road, Lozano slowed the pickup and drove onto another small dirt road through a tunnel of trees. This time, the truck crept over the potholes and loose gravel, and I realized that beyond the edge of the road was a drop-off into a two-hundred-foot ravine. A wave of dizziness passed through me. Lozano built his breeding center in a remote area with strong security fences in an attempt to keep the frogs safe from traffickers.

During our trip, Lozano told me about his life. He first fell in love with animals when he was just four years old. He voraciously read encyclopedias—"there was no internet at that time!" he joked—and he memorized scientific names and facts about all types of animals. In college, he studied zoology but quickly discovered that the program was entirely focused on dairy and meat production. Lozano passionately wanted to learn about undomesticated animals and how to protect them. He was aware of the accelerating rates of extinction resulting from deforestation, ranching, development, and mining, and he could see that no one was thinking about the welfare of wild animals. He went to the dean and explained his budding concept of conservation, and because it was before sustainability and wild animal protection had come into the global conversation, the dean scoffed at the proposal. At that point in the conversation, Lozano glanced at me with twinkly eyes and laughed heartily. He had clearly not taken "no" for an answer.

Undaunted, and fueled by his love of animals, Lozano moved to the United Kingdom in 1999 to obtain a special degree in endangered species management and then furthered his education in animal behavior in Austria. When he returned to Colombia, he became a lecturer at the very university that had questioned his interests—and he began to serve as a conservation consultant.

"The mindset of seeing animals only as livestock was hard to change," Lozano recalled. "The old belief," he said, "was to destroy everything and create something man made." This struck me as a rather common symptom of human hubris, the fallacy that dominion over the earth was a noble goal. This fallacy has guided Western cultures for centuries. At an animal welfare conference in 2004, he made a passionate plea to the attendees. "There are economic benefits to not destroying everything," he proclaimed. "Conservation in Colombia is crucial, in fact. For human survival in the tropics, you need healthy soil, healthy water, healthy animals, healthy insects, healthy trees." Changes needed to be made on so many levels.

Eventually Lozano took a position in Bogotá with the Bogotá District Secretary of Environment, where he was in charge of a department dedicated to wildlife rescue. The animals he rescued were primarily wild animal victims of the illegal "exotic pet" trade. Animal trafficking was a huge problem in Colombia. Countless turtles, capybaras, ocelots, peccaries, snakes, monkeys, and parrots of all kinds were frequently rescued at the airport as they were being shipped out in poorly ventilated crates and suitcases. Others escaped confinement and were found. The animals often suffered from grotesque injuries or malnourishment. Most of them could not be returned to the wild. Lozano and his well-trained crew aimed to keep them alive, particularly birds and mammals. He thought he had seen everything.

One Sunday evening, however, Lozano received a call to come out to examine some cargo. When he arrived at the airport, he opened up a suitcase packed with hundreds of poison dart frogs. As he told the story, he shook his head, his brow furrowing, his mouth turned down in a combination of sadness and disgust: "They were dying so quickly." He momentarily took his hands off the wheel, palms upturned and fingers wide. He spoke as he stared into them. "I was losing them so fast, like sand through my fingers. And at that time, we had no idea how to rehabilitate amphibians. There was no way to save them."

The suitcase full of frogs marked a turning point for him. In addition to the terrible number of animals being stolen from their habitats as part of illegal wildlife trafficking, there was another issue. Those very habitats around Colombia were being destroyed, and if people didn't begin caring, there would be no end to this trend. Extinction numbers had only grown more alarming. Lozano envisioned two solutions, which he would begin to build simultaneously: a sanctuary for rescued animals that would educate children and a breeding center for poison dart frogs. These were big dreams.

When I first climbed into Lozano's pickup truck, I noticed my own anxiety rising and took an intentional, deep breath. Colombia had long been known as one of the most dangerous places in the world, with a fifty-year-long civil war and drug trafficking. In 2016, a peace accord was signed, and overall, the country has seen a huge reduction of violence. But I also knew that despite the peace accord, hundreds of activists—environmentalists and those working for social justice—have disappeared or been killed. Lozano had received threats from traffickers himself. His warm laugh brought me back quickly to the magic of the moment. I felt very lucky to have a chance to see how his dreams had become realities.

Not much farther down the road, we came to a driveway blocked by a huge metal gate. Lozano jumped out and unlocked it. These precious little beings did require serious security. The truck climbed up an even steeper, bumpier road. In the late angle of light, the mountainside was a snarl of trees and plants, and the word "wilderness" came into my head. We have much debate these days about whether wilderness can exist in a world so dominated by humans, in a time we are calling the Anthropocene. But this place did feel "wild," despite the road. "Home to 150 vertebrates and countless insects," Lozano announced. "Six hectares, designated for preservation." We stepped out of the truck and were engulfed in the sounds of insects, birds, and an occasional monkey calling. I bathed in the sound, the sonorous song of a place where so many creatures were still thriving.

Lozano led me down a footpath to a couple of small buildings. Before entering the actual breeding center, we sanitized our shoes, and I was told not to touch anything. We were not to bring any unwanted disease or bacteria into the room of baby frogs.

Inside the small building stood wall-to-wall aisles of illuminated plexiglass containers. Each held a collection of leaves and stones, small water features . . . and a multitude of beautiful frogs. "Ranas de muchas colores!" The many colors and designs represented different morphs, black ones with yellow spots, royal blue ones that looked like they had been splattered with gray paint, bright red frogs with black eyes, gold frogs. . . . And these could kill us, I thought.

As it turns out, these frogs could not kill us. The toxicity of poison dart frogs is entirely dependent on their diets. Poison dart frogs evolved to be able to consume wildly toxic insects, like specific ants in tropical forests. Without this component in their diet, the frogs' skin would not secrete the same toxins. Lozano opened a door and a sound like a thousand untuned violins came forth, vibrating in the air around us. The room held at least ten thousand crickets, the basis of these frogs' diet.

These frogs all fell into the category of *Oophaga*, a genus of poison dart frogs, and the breeding of these frogs in the wild is a complicated affair. The female lays her eggs in the leaves of bromeliads and waits for a male to come and fertilize them. She then carries the eggs farther up into the trees and places the eggs in several different tiny pools of water collected in other cup-shaped leaves. Each day she returns to a different cache of eggs and feeds the tadpoles an unfertilized egg. "This goes on for three months!" Lozano exclaimed. Deforestation devastated frog populations. Stealing one female from her habitat destroyed dozens of future frogs. Lozano and his team mastered breeding in captivity, and after years of applying for the appropriate licenses to sell the frogs and return them to the wild, the government finally said yes.

"Seems odd that it would be so hard to get a license to do something so good for Colombia," I said, curious.

"Don't get me started about all the red tape!" Lozano laughed. "That could be an entire book!! A real comedy!" He said that before they had a license to return the frogs to their habitats, they would joke about dressing up in camouflage to sneak into the forest, not to steal anything but to put the frogs back. He was gut-laughing now. "It's been quite a journey. . . . It's the reason I race cars!"

Today, he sells his frogs and butterflies, legally, to pet stores around the world in order to undercut the illegal market and to replenish populations in the wild. The profits from Tesoros were used, in part, to hire local people living in deforested regions that had once been frog habitats to plant new trees and native plants. "We need to protect our forests. Some of them are in danger now that there is peace, ironically," he muttered, "because before it was too dangerous to take trees out. So the people need to be motivated to protect their land. They have to know what is at stake." This is where the second part of his dream came in: the Bioparque La Reserva, a haven for misfit animals, those who had been damaged by humans.

The Bioparque grew out of Lozano's work with exotic wildlife rescue and currently has about thirty-five thousand visitors each year. The Bioparque housed the animals rescued from the airports that could not be returned to the wild. Instead of being kept in small cages, they now lived in larger, natural spaces on land Lozano had obtained. The sanctuary is dedicated to educating Colombian children about the beauty and mystery of the nonhuman life around them. He hopes to instill in all Colombians a desire to care about their environment. I was astonished at the range of animals. A friendly macaw pulled at my shoelaces. "How do you know how to care for all of them?" I asked.

He stared at me, puzzled at why I would ask such a silly question. "You have to listen to the animals. They will tell you what they need. Humans need to listen."

Darkness descended on the mountain. We all huddled in the meager porchlight outside the breeding center, basking in the music of the night. So many voices singing, a rare and beautiful music I hoped would continue for years to come.

A few days later, I walked with a graffiti artist named "Dave" through the streets of Bogotá. One mural had an Indigenous woman standing in front of a pattern of what appeared to be pineapples but were, in fact, grenades with pineapple leaves emerging from the top. Were these a symbol of the explosive sweetness of Bogotá? Dave asked rhetorically. No, they symbolized the fact that the rich, fertile land of Colombia had become a source of war.

Again, this point was driven home to me. The land itself, a paradise perfect for growing things humans love—coffee, fruit, chocolate, orchids, and cocaine—had been turned into terrifying terrain by human greed and violence. So much life, both human and nonhuman, had been destroyed over time.

Down a narrow brick street, another mural showed a clear-eyed young woman from the waist up, pulling the skin of her chest back to expose a ribcage and a heart surrounded by voluptuous blossoms. The new beginning. The youth were brave enough to show their hearts, to rebuild, to begin this story again with the tenacity, courage, and vulnerability that is love.

So many of the graffiti artists borrowed from Indigenous people's stories. Images of fierce women were entwined with images of jaguars, ocelots, butterflies, and snakes. The power of the natural world, the invitation to reconnection, was everywhere. Behind one large brick building featuring a mural made by a group of artists, honeycomb ribboned through the scenes, filled with honey. Sweetness was available, tangible, if Colombians and the powers of nature came together. Hope with a fist in the air. Above us a flock of pigeons arose in a clatter. Onward.

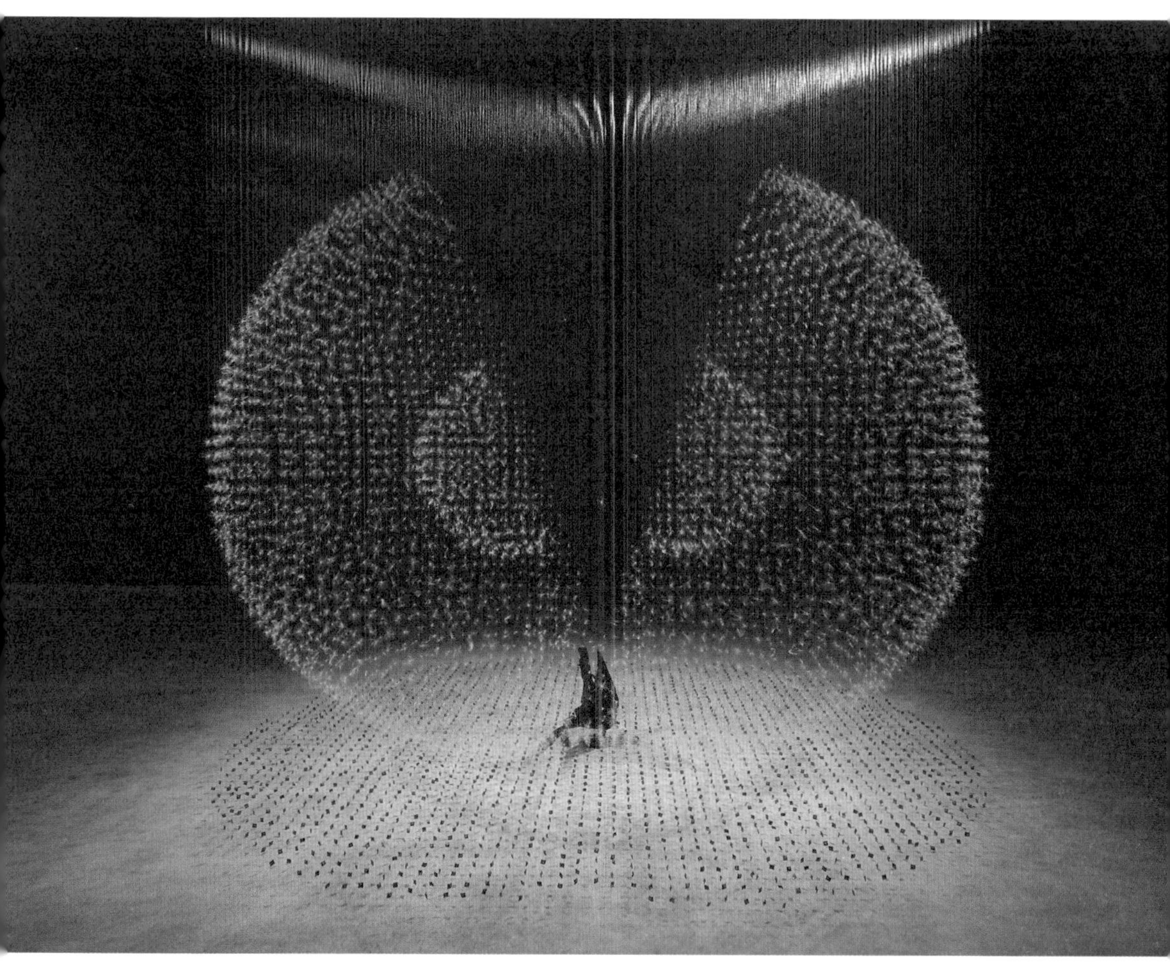

Claire Morgan, *Gone to Seed*, 2011. Carrion crow (taxidermy), thistle seeds, nylon, lead, 300 × 240 × 180 cm. © Claire Morgan. Photo: Jordan Hutchings.

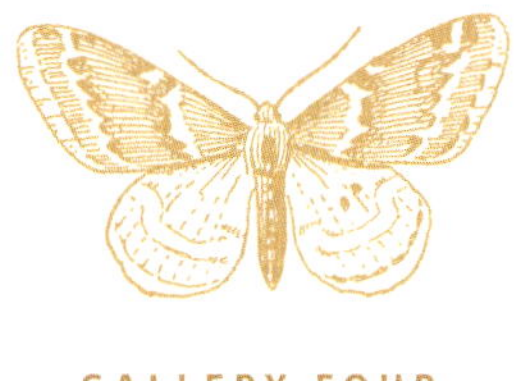

Claire Morgan

Dead Owls and Bluebottle Flies

When you enter a room designed by Claire Morgan, you are immediately transported to a magical place that unsettles your basic human understandings—a floating world that defies gravity and rational thought. The body of a dead fox hovers, midair, surrounded by a precisely measured cube made of suspended bits of ripped plastic. A sphere of milkweed seeds wraps a diaphanous force field around a falling crow that seems to be frozen in time, endlessly falling and also not falling. A plane of bluebottle flies floats above a dead peacock as if carrying his spirit upward. The world Morgan creates is at once beautiful—in its composition, its airiness, its patterns—and wildly discomfiting. Death and life intertwine; gravity and time stop; and pests are presented as lovely, as an integral part of the constellation of beings. Here, you enter the deeper mystery of interconnection between the human and nonhuman worlds in this moment of global environmental devastation.

Many of Morgan's installations feature a central taxidermied animal surrounded by an atmosphere created by dead insects or flower petals or bits of plastic hung on invisible threads. In *Here Is the End of All Things*, Morgan has created three cubes out of suspended thistle seeds and bluebottle flies. The cubes appear to have been tunneled through by the flight of a barn owl, which appears at edge of the last cube with wings open, as if caught mid-flight, perhaps in the moment of exiting this realm. The dead owl looks strangely alive, immortal.

I am not accustomed to picking up newly dead animals, like Morgan does. I do have piles of individual bird feathers and nests, shells, and even the skull of a racoon on my mantel at home, but I have never done the work of picking up an animal who had recently died.

On a canoe trip this summer, around the edge of a marsh, I noticed an awkward splay of long, mottled feathers lying still in a tuft of grasses on a tiny island no more than six feet wide. My friend and I moved our canoe closer to investigate. Face down, with wings spread wide and head pressed to the ground, a great horned owl seemed to have fallen from the sky like a crashing plane. My chest tightened. The ungraceful end of such majesty seemed wrong somehow. There was no sign of violence. What had happened? The soft small feathers by her ears moved slightly in the night air. We could not tarry too much longer or it would be dark. I could not stop thinking about the owl. In a week's time, I went back, and a sprinkle of tiny blue flowers had bloomed on the miniscule island. Forget-me-nots. I vowed to return for the skeleton later, once the insects and time had done their work. Unlike Morgan, I had no idea how to preserve the owl, but I wanted to honor her somehow.

I knew little about the practices of dead animal preservation, but I teach on a campus that has a remarkable zoological collection. Inside one of our large stone edifices are boxes filled with mammal and reptile skeletons, shelves covered with glass jars full of dead fish and amphibians, and drawers filled with extinct birds and the hides of common rodents. "Bones and skins," our curator once explained to me during a visit, "are useful for different kinds of research." It's like a library of dead bodies, nonhuman ones. A few of the skeletons are articulated: a six-foot-long sea turtle from the Galapagos, a chimpanzee, a bat. And several other animals have been preserved to look like they are perched or poised on pedestals: a wolf, a beaver, a great horned owl.

Many of these animals are endangered; some are on the extinction list. Some are already gone. All of these zoological specimens died naturally. Dead animals can tell us a lot about what is killing them. Some have been shot, others sickened by diseases, and others poisoned. They can also tell us about environmental change over time. At the Field Museum in Chicago, the amount of soot found in bird feathers told a story about air pollution over the Rust Belt in the early 1900s. There are so many stories.

In the "skins" preservation room, I watched a woman turn a warbler inside out, slicing the belly first, emptying the body, peeling the skin back carefully so as not to damage the feathers, and then preparing it to be stuffed with cotton. The "bones" preparation involved a flesh-eating beetle colony. The curator and I walked down a musty cement tunnel into a chamber with several large tubs. The beetles would not eat live flesh, she assured me, so we were not in any danger.

When I found the owl, I knew I could not do any of that, so I thought I would let nature do its work, the insects and the fungi so efficient at helping things decompose and return to soil. I assumed that the bones would remain, and I could take the skeleton then.

The great horned owl was not a common sighting to us. Lately, I have been watching juvenile barred owls in the woods near my home. Their already huge wings are nearly silent as they drop to the ground to hunt. Several times, locked in the gaze of their large, piercing eyes—eyes that could spot prey scores of feet from the ground—I felt relieved I wasn't a rabbit, a shrew, or a mouse. Barred owls and great horned owls eat many of the same things, but the great horned owl is the larger apex predator. A

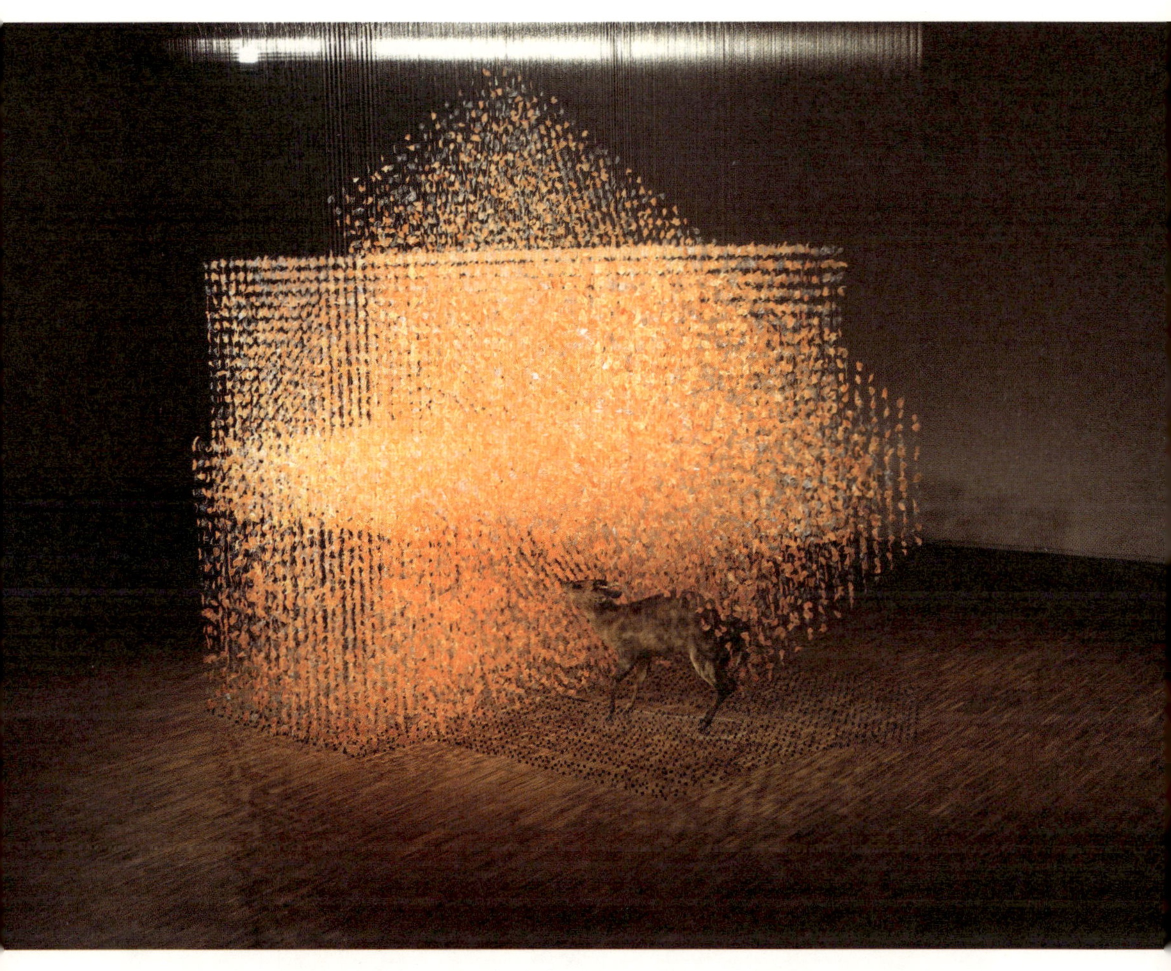

Claire Morgan, *If you go down to the woods today*, 2014. Muntjac deer (taxidermy), butterflies, polythene, nylon, lead, 300 × 300 × 250 cm. © Claire Morgan. Photo: Claire Morgan Studio.

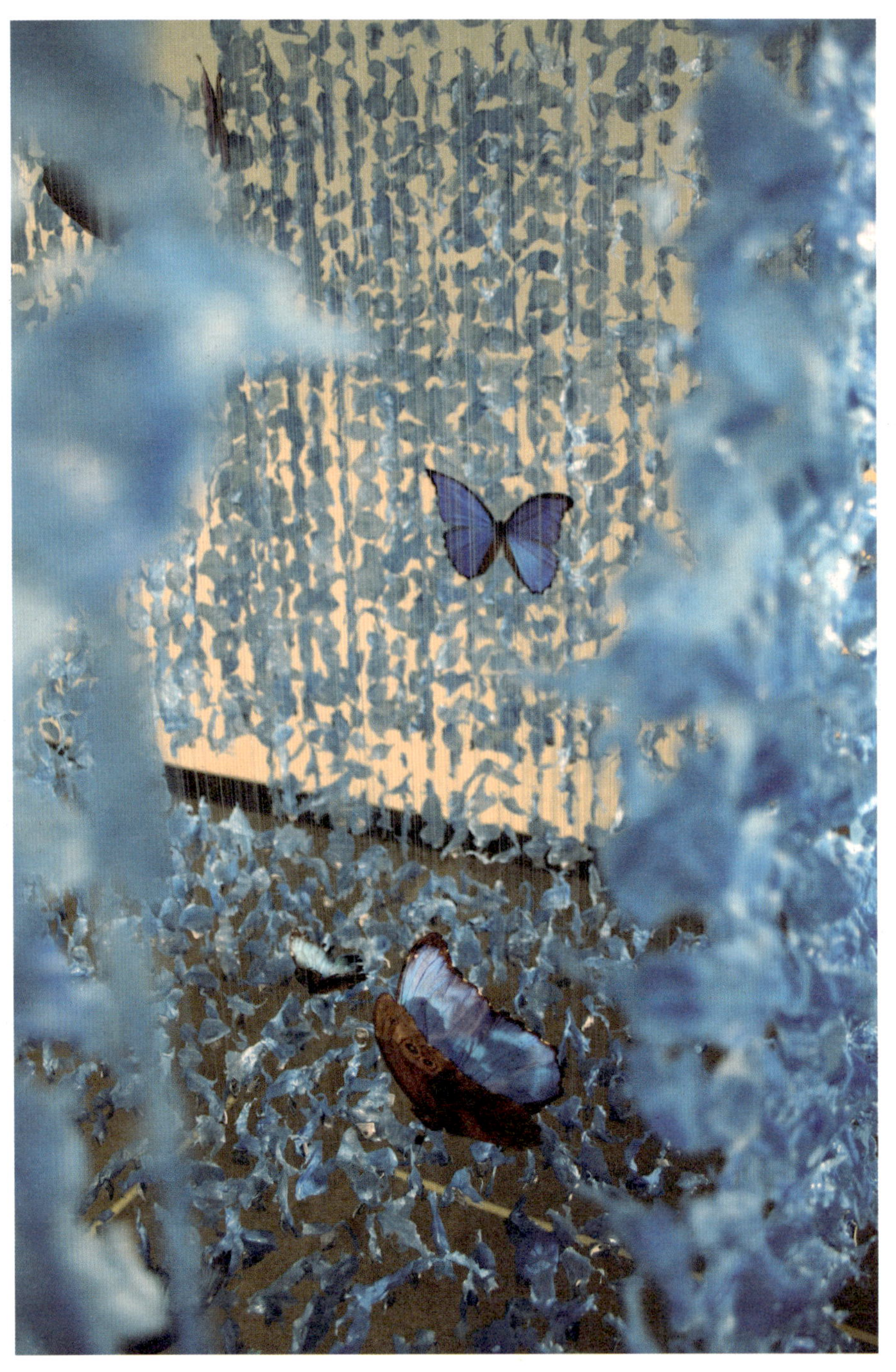

Claire Morgan, *The Blues (II)*, 2009, detail. Morpho butterflies, polythene, nylon, lead, 200 × 93 × 93 cm. © Claire Morgan. Photo: Claire Morgan Studio.

young barred owl could be attacked by a great horned owl, but the great horned owl has few enemies here. We described the death scene to a falconer we knew, and she said the death was most likely due to pesticide poisoning.

The National Institutes of Health defines pesticides as "chemical substances used to prevent, destroy, repel or mitigate any pest ranging from insects (i.e., insecticides), rodents (i.e., rodenticides) and weeds (herbicides) to microorganisms (i.e., algicides, fungicides or bactericides). . . . Over 1 billion pounds of pesticides are used in the United States (US) each year and approximately 5.6 billion pounds are used worldwide." Merriam-Webster's online dictionary defines a pest this way: "a plant or animal detrimental to humans or human concerns (such as agriculture or livestock production)."

To Morgan, this very definition is problematic. She told me, "Whenever I am making my work, I am using bodies of animals, and quite often these are animals that humans regard as pests. . . . We are animals, we are part of the natural world. We tend to behave as if we live ON the natural world, like we are separate entities that can use it, take advantage of it, that it is our resource, when really the earth is one huge organism that we are part of."

The label "pest" gives us permission to kill creatures within the same system we are a part of without any sense of guilt. Our stories and myths have demonized so many creatures: spiders, bats, mice, crows, wasps, snakes, even dandelions. In our attempts to eliminate these beings for our convenience and comfort, we not only break strands of a fragile web but destroy countless other beings in the process. Bees, frogs, songbirds, owls . . . even humans can be damaged by this killing practice. The Audubon Society reports on the horrific deaths of owls who have died from internal hemorrhaging after eating rodenticides. Many others are sickening slowly.

Morgan's work tells us a new story. Her pieces honor all of the animals in them. Pests are just creatures that are mortal, like we are. They are part of our interconnected world. They are kin, and we have created this imbalance.

In one installation, called *Elephant in the Room*, a nebula of tiny pieces of torn paper (that she sees as a symbol of our obsession with reckless consumption) creates the shape of a fourteen-meter-long northern right whale. The enormity of the sculpture is balanced by the immaterial, ethereal aspect of the giant mammal as it swims above you through the air. Ideas of impermanence and the danger of a looming extinction come into my head, but the piece also offers another more haunting sensation, as if this might be a sacred moment caught by an artist, the whale spirit living on. A ghost whale, perhaps with a warning.

The poet and artist Ian Boyden once told me a story about traveling to Mitla, an archeological site in the Yucatán in Mexico. As he walked along the edge of one of the elaborately carved stone buildings, he was suddenly surrounded by thousands of white moths fluttering madly around him. For a moment he was blinded and also completely dazzled by the frenzy of wings. Moments later they disappeared, and Boyden noticed that a guard who stood nearby had seen what had happened and was laughing heartily. "You looked like a cloud," the guard said. "The clouds are the intermediaries of the gods. Looks like they were trying to take you!" Reflecting back, Boyden said that at that time he was "too heavy." He had "no spiritual helium." They had to put him back down.

The whale in Morgan's work seems to be light enough. I think of all the animals facing extinction and how none of them have caused this crisis. I carry the weight of that.

The baby barred owls will be hunting farther from their home soon, over yards covered in pesticides. The other night I sat on the ground in the dark under a tree where one had perched. I watched her swoop from branch to branch, her elegant brown and white wings riding the breeze. She would pause to peer at the ground intently, often looking directly at me, perplexed, perhaps, by my rapt attention. Suddenly, she dropped down to the ground just a few feet from me, staring straight into my eyes. I held my breath, heart beating wildly at this gift. Then, just as suddenly, she took to the air again, the wind from her wings whooshing over my face. How could I do anything but tell her story, the story of her radical trust, her faith that this human would do her no harm?

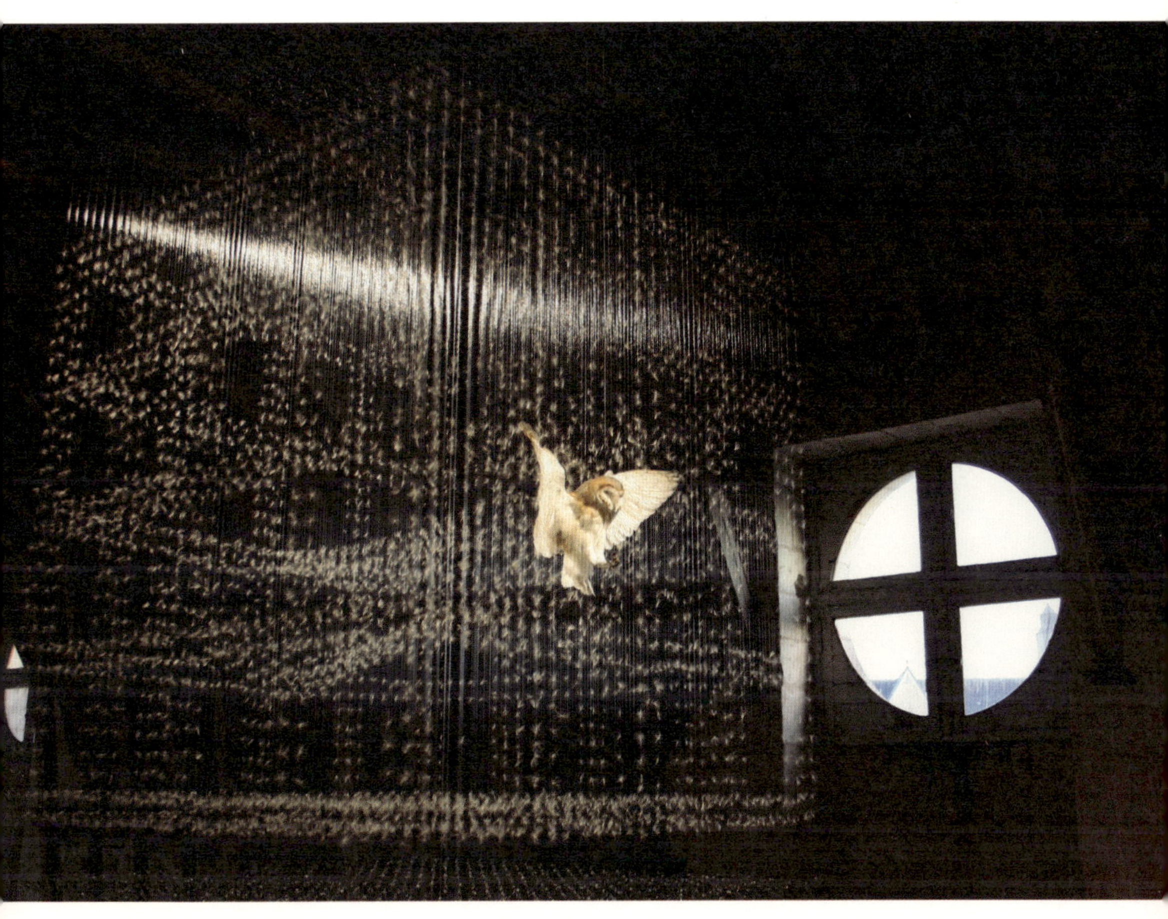

Claire Morgan, *Here Is the End of All Things*, 2011. Barn owl (taxidermy), thistle seeds, bluebottles, nylon, lead, 240 × 900 × 150 cm. © Claire Morgan. Photo: Claire Morgan Studio.

Transformers

BEETLES CHANGING HISTORY

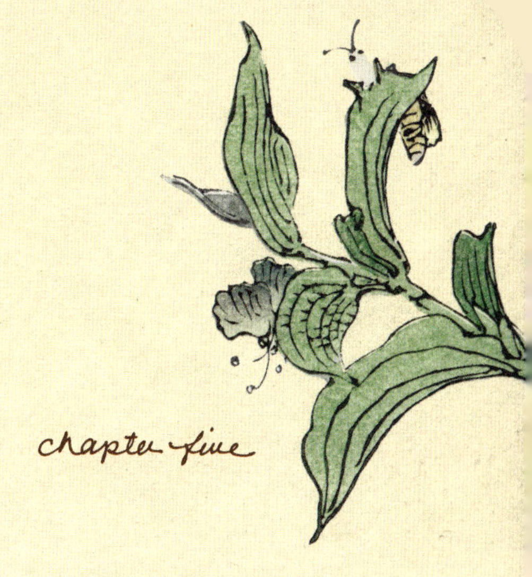

So much of the extraordinary can only be seen by slowing down, by paying close attention, and by looking deeply. Sometimes that deep looking is metaphorical; sometimes it means looking into the deep past. But in the case of forests, it is literal depth that I am thinking about. When we walk through a healthy forest, our feet are stepping on, and past, thousands of invisible species churning through the forest floor to keep the energy of this phenomenal ecological network vital. Fungal communities, microbes, salamanders, earthworms, tree roots, and thousands of insects live just below the surface, doing the essential work of decomposing dead things in order to return minerals and nutrients to the soil so that the cycle of life continues. Beetles, along with fungi, are some of the most essential characters in this story of the underground. I knew little about these elusive creatures, so I sought out the entomologist Michael Ulyshen, who is an expert at seeing what was invisible to me.

I admit my fascination with beetles began much earlier, with scarabs. I had the luck to find a large stag beetle in my friend's yard in Wisconsin a few years back. It had the color of highly polished walnut and long segmented legs, which swam slowly in the air as I held it up for a photo. It looked vaguely steampunk in its design, clunky and industrial somehow, but in truth simply a marvel of nature! The complexity filled me with awe. What was life like for a stag beetle? I wondered. I knew a fair bit about honeybees, but so little about beetles. Egyptians had revered them so much that they made them into jewelry, painted images of them, buried them with the dead. To them the dung beetle represented immortality. Beetles all over the world are famously beautiful. My curiosity about them led me to exchange emails with the "Ministry of Beetles" in Taiwan. Two scientists, Uitsiann Ong and Takaharu Hattori, founders of the group, had produced a book called *Jewel Beetles of Taiwan*, celebrating the vast array of colorful beetles found in the countryside of Taiwan. They offered to take me on a beetle tour if I could make it to Taipei. Sadly, the COVID-19 pandemic kept me from that experience during the writing of this book, but I hope one day to make that trip. Luckily, I did have the chance to meet more beetles in the United States and to learn why they are key to a healthy forest.

My desire to understand the importance of forest beetles expanded when I read Paul Bogard's *The Ground Beneath Us*. In the book, Bogard makes it clear that soil is something we need to be thinking about. He writes,

> Dirt—or more precisely, soil—is quite literally the foundation for human life on Earth. Yet around the world, we are degrading and depleting soil at a reckless rate. One recent study estimates that if we continue our current pace of soil abuse, we have only sixty years of harvests left. Our lack of attention to soil is hard to overstate—there is no human life without it, we are in the process of wasting it, and almost no one aside from specialists seems too concerned. It turns out the soil—the living entity beneath our feet—is the most amazing world that we know almost nothing about.

As Bogard points out, lawns, industrial agriculture, and asphalt are big culprits in soil destruction. Good topsoil is only several inches deep. When we cover it in herbicides, pesticides, and cement, the many thousands of living things that make that topsoil good topsoil are destroyed. Deforestation is another problem. When forests are cleared, soil erodes. Trees help keep soil intact, and healthy soil is necessary not only for a healthy forest but also for a healthy planet. I was curious about how insects contributed to keeping soil healthy. When I learned that beetles who break down decaying wood were critical players, I made a trip to the southeastern part of the United States.

Mike Ulyshen works for the Forest Service and for the University of Georgia at Athens. Ulyshen is a treelike, long-limbed man with gentle dark eyes. His love of insects began when he was just a young boy. When his father had to build Ulyshen's older sister a bug collection kit for a school project, he decided to make one for Ulyshen as well—a blue box, a net, and some pins. Their home in Ohio rested on forty acres of land, much of it retired cropland and parts of it wooded.

The landscape proved to be rich in insects waiting to be discovered by the young entomologist. Butterflies and moths studded his initial collection, but then he became interested in beetles. Their adaptation, diversity, and survival strategies fascinated him, and now his career is partly built upon studying their unique qualities and roles.

Ulyshen and I met at the Oconee Forest Park in the center of Athens, a forest he said was about one hundred years old. Enormous trees towered above us as Ulyshen led me down a red dirt path. As the daughter of a potter, the soil of the path reminded me of earthenware that my mother and I had used in our studio now and then. I mentioned this to Ulyshen. The place we were standing had once been a field of cotton, he said over his shoulder. The good topsoil was all worn away.

I paused for a moment. Of course, this land was once covered with cotton. Much of the South had been. My waves of awareness came slowly. Once this had been a site of slavery, of oppression, of violence. Here, I was, a white woman from a northern city visiting a forest in the South without any immediate personal or historical connection, and I began to imagine all the different lenses one might see this place through.

Once, many years ago now, when teaching a small summer class, I took my students to a local forest for a hike. My students that year happened to be a group of African American men from southern cities who had only recently arrived at my university. I was excited to take them into a place I thought they might find different and relaxing—tall trees, so much green, a contrast to the cities they knew so well. As we hiked, one of them turned to me and said to me, laughing, "This is just like the Underground Railroad." I stopped in my tracks. I realized how foolish I had been to assume that we would have anything close to the same experience of these woods. What a gift this comment was to me. It awakened an awareness of how differently each of us may perceive and experience a landscape. I thanked him, and since that time I have understood very deeply that each of us is born into a particular body with a particular history. These histories, ours and those of our ancestors, and our unique perceptions shape our experience of the natural world, just as the natural world shapes us. How had this landscape and its insects affected the history of the people here and how had they shaped it? I wanted to learn.

The heat index in Athens, Georgia, the day we met at Oconee Forest was 105 degrees Fahrenheit. The warm and humid climate in the southern United States made it perfect for crops like tobacco and cotton. In the 1800s, cotton dominated the economy of the South. In 1793, the invention of Eli Whitney's cotton gin, a machine that removed seeds from cotton, would exponentially expand both cotton production and, tragically, slavery. About 1.8 million enslaved people are estimated to have worked on cotton plantations. Ulyshen said that cotton really decimated the landscape of Georgia. This forest, though, looked pretty healthy to me. I was slightly baffled about how it had transformed from cotton field to forest.

"The oldest trees in the Oconee have been growing here since the 1920s, after boll weevils took over the cotton plantations," Ulyshen said.

"Boll weevils?" I asked, trying to bring back this fact from my history classes. I had come to see beetles but had not expected to discover that one particular beetle had utterly changed history.

One thing I knew about weevils was from a trip I had taken to northern Wisconsin, where I spent some time exploring bogs with a botanist named Susan Knight. She told me about an interesting native water weevil that ate Eurasian milfoil, an invasive plant plaguing many northern lakes. Boll weevils, I learned, had a much bigger impact. "In 1892, the boll weevil came in on the wind from Mexico and destroyed all the cotton," Ulyshen told me as he strolled ahead. "One insect decimated a billion-dollar industry." Did I sense some ambivalence in his tone?

The boll weevil is a small gray-brown insect with a body resembling a kernel of corn with a long snout like an elephant. It is not a fast-moving insect, but one female can produce two hundred eggs. It blew into the United States and nestled in the unopened bulbs of the cotton plants. The voracious larvae ate everything inside. Within a few years, the dominant crop of the South, King Cotton, was down by 50 percent. This was obviously devastating to those who counted on cotton for their livelihoods.

In an article on North Carolina State University's website, a report by the entomologist Dominic Reisig called "The Boll Weevil War, or How Farmers and Scientists Saved Cotton in the South" narrates the momentous efforts waged against the boll weevil. In 1903, the chief of the federal Department of Agriculture called the growing insect population a "wave of evil." The weevils, who were not native (and lived on cotton, a plant that also was not native, having been introduced by colonizers), had no natural predators. During the next fifty-odd years, farmers poured copious amounts of pesticides on their fields. According to Reisig, "During the 1950s, controlling boll weevil infestations required multiple applications of very harsh and toxic insecticides (e.g., aldrin, azinphosmethyl, benzene hexachloride, chlordane, dieldrin, toxaphene, malathion, methyl parathion, and parathion)." DDT was another popular pesticide. The weevil built resistance to pesticide after pesticide, and the constant rain of toxins had deleterious effects on the water, the animals, and the humans in its wake.

I couldn't help thinking about how often humans choose to use incredibly toxic substances to eliminate something we have deemed a "pest," often without foresight. A short story by T. C. Boyle called "Top of the Food Chain" comes to mind. In the story, the audience hears just one side of a defense from a character who was involved in spraying a town in Borneo with DDT only to see a cascade of disasters. The lizards that eat the worms that damage the leaf roofs of the inhabitants' homes are dying, so roofs are collapsing, and the cats that eat those lizards are dying, too, all due to the bioaccumulation of the pesticide. Rats then become rampant. Eventually cats from other places are airdropped in to try to save the day. The narrator is basically shrugging his shoulders, without much conviction, saying, "How could we have known?"

Often, though, we do know the deleterious effects of pesticides, but we make these decisions to use them anyway. Parkinson's disease, the illness my father struggled with for years, is clearly linked to pesticide use. Lymphatic cancer is clearly linked to pesticide use. The decline in bees is clearly linked to pesticide use. And Rachel Carson warned us about bioaccumulation as early as 1962 in her seminal book, *Silent Spring*. Still we pour and pour these toxins. It was not a surprise to me to read that at one point in history, one-third of all the pesticides used in the United States was used to eradicate the boll weevil.

However, some historians and economists have seen the boll weevil as something of a boon. Historians Richard B. Baker, John Blanchette, and Katherine Eriksson revealed the impact of the boll weevil on education, for example. Since cotton was a "child-labor-intensive crop," a downturn in cotton production allowed children more time to go to school: "Both white and black children who were young (ages 4 to 9) when the weevil arrived saw increased educational attainment by 0.24 to 0.36 years." A working paper from the National Bureau of Economic Research, "When Coercive Economies Fail," examines the social impact of the decline of cotton. The authors provide evidence that the shock to the institution of cotton production resulted in a less violent South, with fewer lynchings and less Ku Klux Klan activity. Some historians even credit the boll weevil, in part, for the Great Migration, a period when millions of African American families and individuals moved north for better lives, introducing a period of rich literary and intellectual expansion.

The song "Boll Weevil" evolved from "Mississippi Boweavil Blues," a piece originally written by one of the founders of the Delta blues, Charley Patton, in 1929, a time when the boll weevils' devastation was clearly being felt by cotton plantation owners and the many tenant farmers who worked for them. The entomologist Robert K. D. Peterson wrote an extensive history of Patton and "Mississippi Boweavil Blues":

> Patton may have been performing this tune as early as 1910, making it among the earliest of blues songs. By that time, the boll weevil would have been known to nearly everyone in the deep South. Since entering the United States near Brownsville, Texas in 1892, it moved rapidly over the next 30 years to infest 600,000 square miles. In 1907, the first boll weevil was found in Mississippi, and by 1915 populations covered the entire state. The social and economic upheaval wrought by the boll weevil is legendary. Cotton crops failed, land prices sank, farmers went bankrupt, and farm workers fled *en masse* to industrial cities in the north.

Peterson argues that the boll weevil in the song is clearly a threat to plantation owners and tenant farmers—but also a source of envy to the tenant farmers who did not have the weevil's mobility and power. The cotton plantation workers faced economic instability due to the weevil, and moving North was not as simple for humans as it was for this insect.

Peterson points out that the music critic Ayana Smith thought of Patton's boll weevil as a trickster, a figure in African American narrative tradition that "flouts the norms of society, using cunning, humor and deceit to obtain personal gain." The song is a powerful metaphor for African Americans who were also "looking for a home" but did not enjoy the same freedoms of movement that the boll weevil (or white Americans) did. One version of "Boll Weevil" features a conversation between a narrator and a boll weevil who is destroying the crops wherever he goes:

> Sees a little bo weevil keeps movin'
> in the air, Lordie!
> You can plant your cotton

and you won't get half a bale, Lordie
Bo weevil, bo weevil, where's your native
home, Lordie
"A-Louisiana raised in Texas,
least is where I was bred and born," Lordie
Well I saw the bo weevil, Lord,
a-circle, Lord, in the air, Lordie
The next time I seed him, Lord,
he had his family there, Lordie
Bo weevil left Texas, Lord, he bid me
"fare ye well," Lordie

(Spoken: Where you goin' now?)
I'm goin' down the Mississippi, gonna give
Louisiana hell, Lordie

(Spoken: How is that, boy?)
Suck all the blossoms and he leave your
hedges square, Lordie
The next time I seed you, you know you had
your family there, Lordie
Bo weevil meet his wife,
"We can sit down on the hill," Lordie
Bo weevil told his wife,
"Let's trade this forty in," Lordie
Bo weevil told his wife, says,
"I believe I may go North," Lordie

In Enterprise, Alabama, there is actually a monument to the boll weevil sculpted by an artist from Italy. The monument, a classical statue of a woman, honors the insect with a plaque reading, "In profound appreciation of the Boll Weevil and what it has done as the Herald of Prosperity." Why would a town in the South be grateful for an insect that was decimating the crop they had depended upon? Because it forced diversification. Farmers in the community began to "heed the advice of such agricultural scientists as Tuskegee Institute's George Washington Carver and diversify their planting to include peanuts, sweet potatoes, and soybeans. By following Carver's advice, Coffee County, in contrast to the rest of the state, rebounded economically in 1917 with the largest peanut harvest in the nation."

In *Freedom Farmers: Agricultural Resistance and The Black Freedom Movement*, Monica White argues that Carver's ingenuity laid the foundation for contemporary permaculture and urban farm movements. She writes, "Corn and cotton, the South's longstanding cash crops, had significantly depleted southern soils." White also observes that "while many agricultural historians and current practitioners and proponents of the local food movement consider J. I. Rodale—who wrote a highly influential book on organic agriculture in the 1940s—to be the father of sustainable agriculture, Rodale's work depended on Carver's. . . . Carver provided evidence of the value of composting, crop rotation, and diversification—three basic tenets of organic farming." The cash crops, like cotton, left behind depleted soils and malnourished tenant farmers and sharecroppers. While many large commercial farmers began using chemical fertilizers, Carver advised the use of cover crops and barnyard waste combined with compost. Would these ideas have had as much purchase if there had not been a rupture in the cotton industry? If this insect had not arrived?

The boll weevil even had a restaurant chain named after it.

There are still many, many people who seek total eradication of the boll weevil. Scientists eventually came up with a way to snare boll weevils using their own pheromones, which would entice them into a trap where they could be killed with pesticides. Cotton is still vulnerable, however. In some states it is illegal to grow cotton without the proper licenses, because it could lead to another swell in the weevil population. There are tight regulations to keep the boll weevil in check.

For Mike Ulyshen, the end of the dominant monocrop simply meant that a forest had a chance to grow. And there we were, that day in Athens, walking along the red dirt path, surrounded by trees that had reclaimed the land: crepe myrtle, southern magnolia, American Beech, loblolly pine, tulip poplars, bald cypress, eastern redbud, and eastern dogwood. How did that happen? The process of succession was slow: new seeds and grasses, bacteria rebuilding, trees, and eventually a new ecosystem. On the sides of the path was a thick layer of leaf matter, and beneath that, soft dirt. Occasionally, huge tree trunks lay flat against the ground. This was the gold mine.

With a small hatchet, Ulyshen gently tapped away at a huge fallen tree, a tulip poplar, already dotted with lichen. The bark had fallen off in places, exposing wood that appeared to still have some moisture in it. He chipped back the intact bark and tore some of the wood below off the tree. He held it up to his nose and breathed in deeply, like he was smelling fine wine. "Aromatic!" he announced, and handed it to me. The phloem made the wood smell fresh and sweet, almost like a honeydew melon. "This is what the insects smell," he said. It had never occurred to me that the inside of the tree would smell so sweet, like a spring flower, that beetles would be attracted to wood the way a bee is attracted to a blossom. I recalled the incredible scent capabilities of a honeybee, thousands of times greater than ours.

I wondered what we would find. Earlier, Ulyshen had shown me a small collection of insects divided into four groups: (1) bark beetles, who ate nutrient and carbohydrate-rich phloem immediately after the tree died; (2) beetles who lived in rotting wood; (3) beetles who lived in collapsing wood, which was much more broken down; (4) and beetles who lived in tree hollows. Each group makes way for the next, and all of them look radically different.

Each of the insect groups had special qualities that made them perfect for helping the tree decompose. The bark and ambrosia beetles compete with fungus for the phloem. Their bodies are flat—"It's like they live in a two-dimensional world," Ulyshen exclaimed—so that they can squeeze under the bark. When we find one, he makes sure to show me a side view, flat as a button. Many of these were very small but have a lasting impact. When you have seen a log with intricate line drawings carved into it, you have probably witnessed the work of a bark beetle. As they do the work of separating the bark from the dead tree, new habitat becomes available for many other insects and spiders.

Next came the insects who lived in the newly dead wood and munched tunnels into the logs, creating space for others behind them. Passalids, large beetles with a long torso and wide head that seemed a bit similar to the stag beetle I had found, were common in these woods. They were important, and Ulyshen hoped we would see some. It didn't take long.

He chopped away at different logs as we worked our way through the forest. As we leaned over one, he said casually, "We have scorpions down here, and sometimes I find them under the bark of these trees. One time I found a mother with a bunch of babies on her back!" His joy over such a discovery made me chuckle. I admit, I was not psyched about having him repeat that experience in my presence; I steeled myself in preparation. We did not end up finding a scorpion, but moments later, Ulyshen opened up a log, and there in the wood was the body of a snake. "Oh, there's a ringneck in here," he said, without a trace of fear. I jumped back as the sleek, thick body of a dark gray snake slithered quickly out of view.

Digging into a decaying oak log, Ulyshen announced, "Here we go: a passalid larva." Passalids are considered subsocial insects because they live in family groups. Under a thick layer of wood was a translucent white wormlike creature, about as long as a quarter. Ulyshen explained that the larvae ate the frass of the parents, thereby recolonizing the microbes in the log, which helped the process of transforming this log into soil.

The charismatic scarab beetles live in wood that has been softened by fungus and bacteria and burrowing insects like the passalids. The most impressive of these are arguably the South American Hercules beetle, *Dynastes hercules*, and the African Goliath beetle, *Goliathus giganteus*. The latter can grow up to ten centimeters long. In his book *Life Everlasting*, Bernd Heinrich points out that African Goliath beetle larvae can weigh "up to 120 grams, about ten times the weight of a warbler." Not everyone would be thrilled about finding a giant bug, but you must admit, that fact is impressive.

Ulyshen is an exception to the rule. Kneeling by another log, he gasped. "Oh! My daughter will be thrilled. She has been hoping to find one of these!" There in his hand was a passalid pupa. His daughter, Ashley, who was eight, wanted to raise one of these beetles. Ulyshen carefully took out a box, added what he hoped were sufficient supplies of wood, and gently placed the pupa inside.

Not all of the dead trees in this forest had fallen. Those dead trees left standing hosted woodpeckers of many varieties, mushrooms with crazy shapes and colors, and small mammals. Ulyshen pointed out that forests and forest populations were healthier if the dead trees were left to decay in the woods. He speculated that perhaps the reason we practiced forestry the way we did was, in part, because of the aesthetics of having an even carpet on the ground or perhaps because of the desire to see what is out there, unimpeded by dead, fallen trees. Depending on where you were in the world in certain moments in history, a wolf or a jaguar could potentially be hiding behind a large log. Overall, though, the dead wood was important for a healthy forest, and our management was not necessarily the best practice for the life of the forest. Ulyshen said that in Europe, particularly in Scandinavia, they were thinking differently about these things.

One group of scientists in Sweden did some groundbreaking research about the importance of veteran trees and the number of species dependent upon these trees. The studies suggest that while replanting forests is a fine idea, the oldest forests are the healthiest. Furthermore, leaving the decaying wood in the forest helps the fungal networks and the insects working the soil on which these trees depend. Old-growth forests are disappearing, and soil is eroding. Beetles are an important part of keeping forests and soil healthy. The studies suggest we need to be thinking about conservation efforts for these crucial insects.

As we walked out of the woods, through the heavy, moist air, it hit me that beetles were entirely responsible for this forest. First, the boll weevil destroyed an industry that was in many ways destructive to land and people, making way for these trees to grow tall. Now, the beetles living in the dead wood were restoring the soil, making it richer so that new things could grow and this forest could be healthy for years to come. Renewal, in this context, seems not only possible but probable. The earth's ability to restart, for life to continue, comforted me. Some lines from the poem "Song of Myself" by Walt Whitman came to mind:

> The smallest sprout shows there is really no death,
> And if ever there was it led forward life, and does not wait at the end to arrest it,
> . . . .
> All goes onward and outward, nothing collapses

If only we were not so relentless in our destruction.

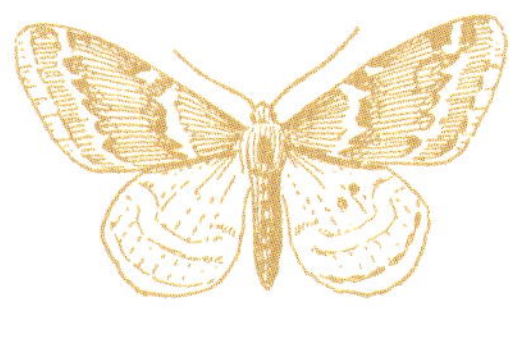

Edouard Martinet

Superheroes

When I first wrote to the French artist Edouard Martinet about his insect sculptures, he sent me some lists of seemingly random objects. The lists included bike spokes, surgical retractors, windshield wiper blades, acetylene light parts, and snail forks, to name a few. All of his remarkable sculptures are assembled from objects he has found in flea markets, thrift shops, and junkyards. Each segment of each piece is carefully screwed together, never soldered, in order to retain a clean aesthetic that will most resemble the creature he is re-creating.

Born in Le Mans, France, in 1963, Martinet attributes his earliest awareness of the magnificent world of insects to his elementary school teacher, who was also an entomologist. In the classroom, students studied and raised insects, and Martinet's love of nature was ignited. After completing a Bachelor of Fine Arts degree from ESAG, an art school in France, and working for a couple of years as a graphic designer, Martinet turned to his true calling, sculpting animals with recycled materials. In his studio are drawers and boxes and piles of scrap metal of all sizes and shapes. Watching him choose pieces and place them together is like watching someone doing a drawing where, at first, there are only shapes and lines, and then suddenly a creature appears. A muffler pipe becomes a torso; a layer of bicycle logos becomes a bank of fish scales; a taillight becomes a grasshopper's eye. Discarded trash is transformed into something elegant and beautiful. What is even more astounding is that the insect sculptures seem animate, filled with personality and energy, rather than like simple machines or monsters. I find delight in this quiet rebuke to Descartes's notion that animals are solely machines. Martinet intends to make his audiences feel something, to shift their awareness. He says, "I try to show people that the animals that could be scary are actually a work of art once you look at them properly. I simply show that nature is perfect."

I admit, when I first saw Martinet's work, I had been watching a lot of superhero movies featuring characters like Iron Man, Ant-Man, Spiderman, Nebula, Dr. Strange, and Shuri, many of whom wear some kind of exoskeleton. Several can fly, and others can shape-shift. Superheroes, I was realizing, had a lot in common with insects. In the recent film adaptation of *Dune*, the most advanced flying machines on the screen look and fly like dragonflies. While many people revile insects, it is clear that part of their reaction comes from fear. The capabilities of insects are simultaneously intimidating and enviable. It makes sense that we would borrow their inhuman traits and give them to our superheroes. And when insects are crawling through a room or flying through the air, it is easy to be afraid. But Martinet's insects are more beautiful than frightening, and they are far more than a bunch of spare parts.

Here are Martinet's lists.

Grasshopper:

Wings: moped chain guards

Abdomen: bike fender, shell-shaped drawer handles

Rear legs: bike forks

Forelegs: bike brakes

Ends of legs: plugs for plaster walls

Thorax and head: pieces of cars and bikes

Eyes: acetylene light parts

Antennae: bike spokes

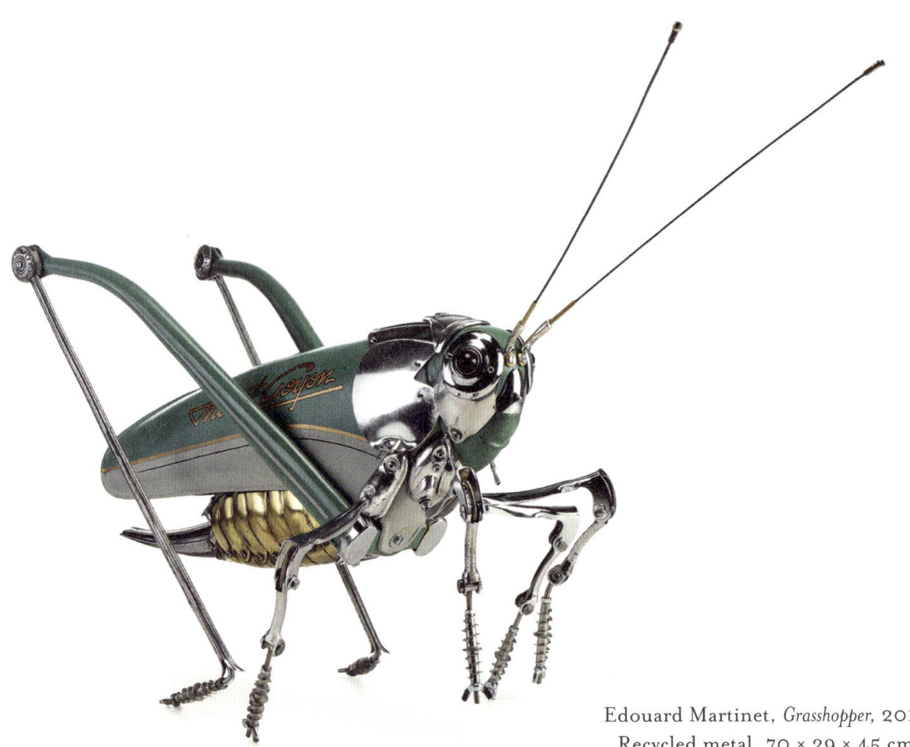

Edouard Martinet, *Grasshopper*, 2012. Recycled metal, 70 × 29 × 45 cm. Photo: Xavier Scheinkmann.

Ant:

Abdomen: bike or motorcycle headlights

Thorax: shoe tree, ball furniture casters, car ornament, bike headlights parts

Head: motorcycle head fork, bike brake parts, typewriter part, pieces of a daisy wheel

Eyes: marbles, upholstery nails

Antennae: bike cable guides, bike brake parts

Legs: cream chargers, brake parts, spoon handles

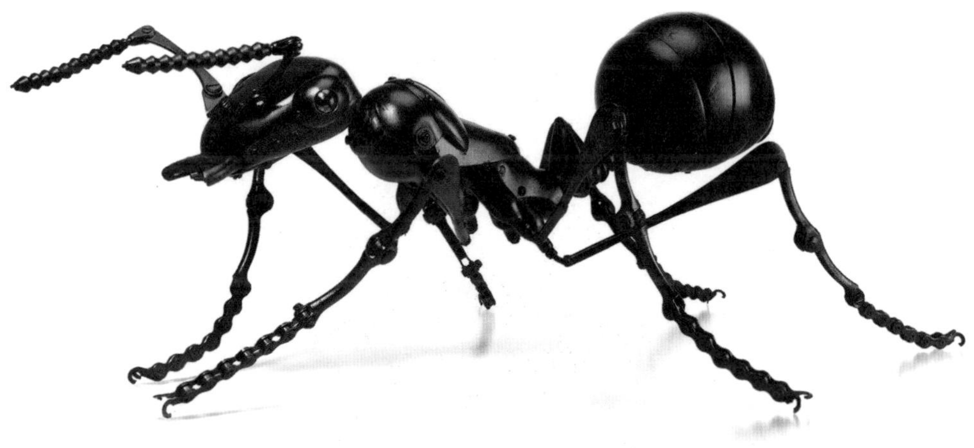

Edouard Martinet, *Ant*, 2016.
56 × 37 × 34 cm. Recycled metal.
Photo: Xavier Scheinkmann.

Fly:

Legs: windshield wiper arms, bike brakes, bike chains, small typewriter parts

Head: motor vehicle rear light

Proboscis: car hood hinge

Antennae: ski boots fasteners

Thorax: motorbike headlight

On the top: 50s kitchen utensil

Wings: the glass is set in a windscreen brush holder, the wing ribs are made with soldering wire

Abdomen: motorbike headlight, part of ceiling lamp

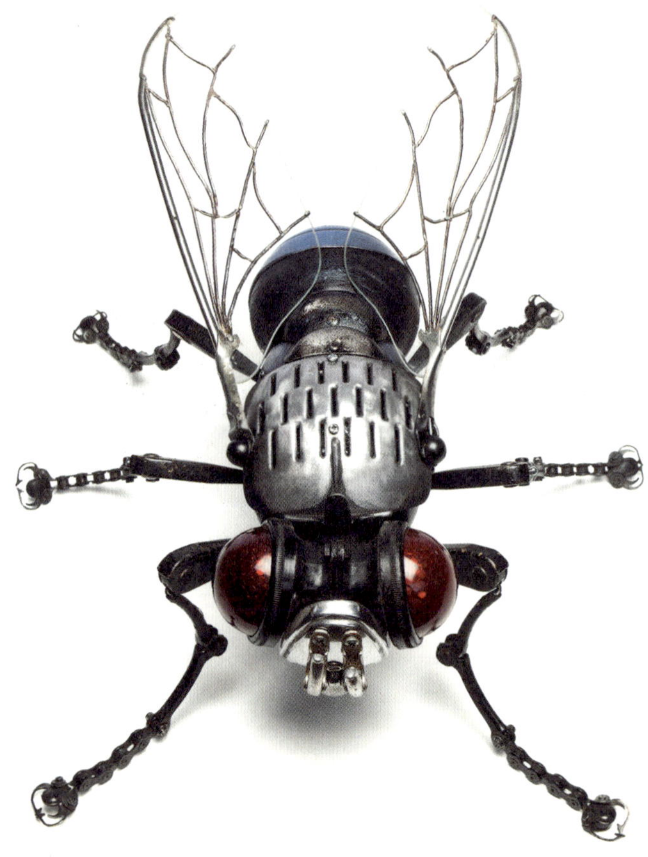

Edouard Martinet, *Fly*, 2016.
Recycled metal. 40 × 47 × 27 cm.
Photo: Xavier Scheinkmann.

Stag beetle:

Head: metal glasses box, rear car lights, glasses parts

Eyes: bike light parts

Claws: universal pliers

Antennae: surgical retractors

Thorax: metal and leather glasses box, rear car lights

Abdomen: Citroën Traction boot handle

Under the abdomen: shell-shaped drawer handles, Solex front fender

Legs: snail forks, bike chain and brake parts, windscreen wipers

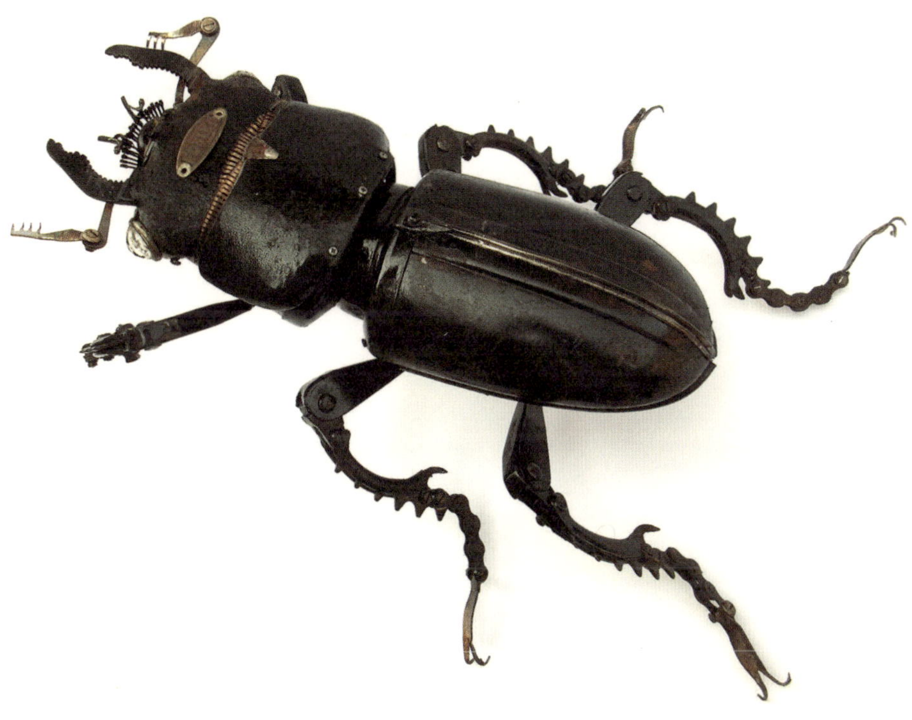

Edouard Martinet, *Stag Beetle*, 2013.
Recycled metal. 46 × 34 × 21 cm.
Photo: Edouard Martinet.

Martinet's recycled scrap metal insects honor the dramatically nonhuman anatomy of specific species. Their larger-than-life sizes emphasize the strangeness of segmented legs, wings, armor-like thoraxes, and enormous eyes. Like Robert Hooke's drawing of a flea, these sculptures surprise with their beauty as well as intimidate us. They look so powerful and so radically different from us. Over time, we have told stories about plagues of insects and made horror movies like *The Swarm*, *Empire of the Ants*, and *Them* in which humans are terrorized by a nature that is out of balance. But why are things so out of balance? Often, it seems, it is the human who has chosen to live in a way that is out of sync with an ecosystem that creates a situation uncomfortable to humans. It also makes sense that the very things we fear are part of the fantasies of being more powerful. Exoskeletons have inspired military uniforms for centuries. From knights' armor to the body armor used by modern police and football players, hard outer shells, if you will, protect the soft, vulnerable body.

But in truth, insects are not as invulnerable as our myths have made them. Edouard Martinet is a hero himself for making use of the abandoned parts we throw away, reminding us to pay attention to the important beings whose futures largely depend on choices we make. Transformations are possible.

In Midewin National Tallgrass Prairie in northern Illinois, I saw an example of how a transformation that seemed impossible was made manifest.

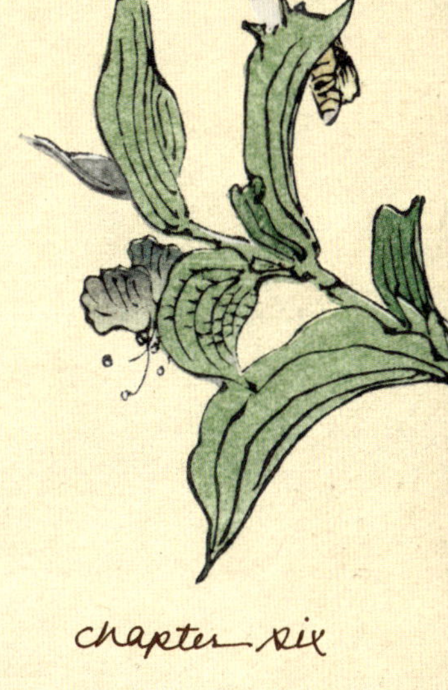

Midewin

ARSENAL AFTERLIFE

> Each moment is an opportunity to make a fresh start.
>
> —PEMA CHÖDRÖN

I often think about whether we can feel hopeful about our planet, and one way I achieve a sense of hope is by finding people doing the important work of change. I need to do this for my children, my students, and myself. But recently, I read a paper called "Agricultural Intensification and Climate Change Are Rapidly Decreasing Insect Biodiversity," and I admit I encountered a setback in my cheerfulness. In the paper (which had seventy-six citations with relevant research), Peter Ravena and David Wagner alert their readers to the fact that we have entered the Sixth Extinction and that insects are disappearing at an alarming rate, in large part due to habitat reduction and pesticide and fertilizer use. They write, "The tallgrass prairie of central North America once extended from Manitoba to northern Texas, covering some 60 million hectares. Less than a tenth of this ecosystem remains; virtually all of the remainder has been given to agriculture." The authors call for radical changes in our farming practices as well as restoration of ecosystems. The book you are reading now is filled with possibilities for change. But what about the really damaged landscapes? Is it possible for them to be restored? We have some very depleted spaces. We have spaces we consider too toxic to live on. How do we solve that?

And then I discovered another story that reminded me that humans are capable and smart, and when there is a dream and people willing to work on it, there is a way. The story is about a place called Midewin National Tallgrass Prairie—on land that was once the site of one of the biggest ammunition plants in the United States during World War II.

The Midewin National Tallgrass Prairie thrives on 18,500 acres given to the US Forest Service after it was decommissioned and assessed to be toxic. At the time it closed, the soil (and water) was not just devoid of much natural life; it was poisonous and labeled a Superfund site. But after the rehabilitation project had been underway for twenty-five years, monarchs used the trees on the oak savanna as a resting point in their migration south; bison herds roamed on the shortgrass pastureland; and Midewin had won a land reuse award from the Environmental Protection Agency. The Forest Service had partnered with The Nature Conservancy, local schools, and the Openlands organization to make such momentous change possible. As a teacher, a parent, and a human being, I had a deep desire to believe that we could heal lands that had been abused, and I wanted to meet the people who made this happen.

On a warm afternoon in September 2021, I pulled off the highway at a spot called Iron Bridge Trail near Joliet, Illinois, just a mile from the Abraham Lincoln National Cemetery. A woman in a Forest Service uniform greeted me with a sunny smile. Veronica Hinke was the public relations representative for Midewin, and we had been in contact for several weeks before my visit to walk the tallgrass prairie. Accompanying us were the botanist Michelle Pearion and Jerry Hoffman, the superintendent of an esteemed fire team, the Eastern National Interagency Hotshot Crew. The three of them would introduce me to the land that was now used for ecological restoration, grazing, conservation education, and recreation.

I wanted to share with them the nature of my work, so I pulled my first nonfiction book out of my satchel. I held up *Where Honeybees Thrive: Stories from the Field* and thought they would perhaps want to look at it. "The honeybee is not native to this country," Hinke stated curtly. Great start, I thought. "I know," I said. Neither is wheat, cotton, sugarcane, or (to my state) rainbow trout. The honeybee had been introduced to this country with the colonizing Europeans. So yes, problematic, as so many things are. My first book looked critically at how the honeybee, an insect I have a deep personal connection to, has been used and abused by industrial agriculture and how those practices, especially intensive farming and pesticides, were not just bad for honeybees but for all pollinators and other creatures as well. It occurs to me that the honeybee is a straw man, an easy target to blame for insect decline. The honeybee may be plentiful, and, at times, competing for forage with native bees and introducing new diseases when they are produced and transported around the country en masse. But this is certainly not the fault of the honeybee herself. The argument that the honeybee is to blame conveniently obfuscates the larger problems, which are industrial agriculture, habitat decline, and ubiquitous pesticide use. I didn't say all of this at the time; I only noted that the first book looked critically at the honeybee's role in agriculture, and the second book was about other insects and their habitats. "I am very interested in native bees and other pollinators for this book," I said, hoping that they were still willing to speak to me, and I tucked my book away. Pearion seemed vaguely unconvinced of my legitimacy, but we agreed to begin the tour.

We passed under a grove of oaks and walked out into a field of flowers and grasses higher than my head. "Just last week this field was covered in monarchs and all kinds of bees," Hinke said. "And moths! For every one blossom pollinated by a butterfly, nine are pollinated by moths," she added. "We just don't notice them because they pollinate at night."

Pearion led the way, plunging through the dense web of stems and leaves. She was born in Illinois, but, like the idealistic adventurer Chris McCandless (made famous in Jon Krakauer's book *Into the Wild*), she craved the north. After high school, she packed up her Jeep Wrangler and drove the very long path to Alaska, where she fell in love with plants and got a degree in botany. She did a stint for the Bureau of Land Management in California but eventually came back to Chicago for a job at the Chicago Botanical Garden, where her passion for prairie plants intensified. A few years ago, a job at Midewin opened up, and now she was doing work she loved most: prairie restoration.

Many of the flowers lacked petals now, standing dramatically silhouetted against the broad sky. Grasshoppers flitted away with our steps. Crickets sang happily. I had read that in mid-August, an acre of healthy tallgrass prairie in the Midwest can support ten million insects. I also remembered a striking study that found that chemically dependent corn fields had next to nothing living in them. Pearion began introducing me to the prairie plants: New England aster (the color of which, Hinke said, reminded her of grape Kool-Aid), gentian, yellow coneflower, monarda, mountain mint, and stiff goldenrod. Each of these plants was home to insects. The goldenrods, for example, attract solitary wasps, soldier beetles, and specialist bees, like mining bees, polyester bees, and the long-horned bee. Plus they were a favorite of fireflies. At all times of the year, different plants offered different things to different insects. We stood in a sea of plants.

I tried to imagine what this place had been seventy years ago, with the arsenal in full production mode. What had that looked like? Over a billion tons of TNT were produced here. Over 926 million bombs, shells, mines, and detonators. Over 23,000 acres had been covered with railroad tracks, bunkers, and arsenal buildings. Over 10,425 people worked the production lines at the Joliet Army Ammunition Plant in the plains of northern Illinois to fill army orders as World War II burned on. And as a result of all this human enterprise, the land absorbed the toxic by-products of war and ammunition production. When the factories were in operation, it was said that "the creeks ran red" from the toxic waste pouring off the site. Soil eroded to bare dirt from the wear of truck tires and boots. This, too, was the by-product of war.

I thought of the educator Andrew Mahlstedt's arresting essay "On Edge in the 'Devil's Gardens' of Bosnia," in which he describes hiking across a mountain slope near the school where he was teaching in postwar Bosnia. A local told him not to step off the trail because the soil was filled with land mines. Mahlstedt was literally standing on the invisible—and still potent—possibility of violence left behind after a war. The potential for violence at the decommissioned arsenal was not due to land mines but was significant nonetheless. Rob Nixon's concept of "slow violence" floated through my consciousness. Nixon coined this term to describe the ways in which environmental degradation and violence against the earth and its inhabitants (in particular the most vulnerable among us) can happen so incrementally that it is imperceptible to our short attention spans. Some kinds of toxicity are invisible to the eye at any timescale. Examples of this are events like the yearslong chemical poisoning of the people of Bhopal, India, after a Union Carbide/Dow chemical plant accident or the radiation that still exists in Fukushima after the nuclear disaster of 2011. As I looked around, I knew that healing the land, at least, was possible in some places. What exactly did that healing entail?

In the earliest days of the recovery project at the Joliet plant, the army was responsible for removing two large contaminated oil pits, detoxifying the soil through treatment or removal, and capping landfills. Phytoremediation, a process that uses plants and plant material to help pull toxins like explosive compounds out of the soil, was also used. Groundwater plumes were monitored to understand where the contaminants were found. Slowly, the earth under my feet had been detoxified. This portion of the site cleanup was completed in 2008.

The creation of this thriving tallgrass prairie required expertise in many areas. Botanists like Pearion knew which plants should be reintroduced. This prairie is meant to re-create what existed here a couple of centuries ago. Long before any Europeans settled on this land, it was inhabited by several groups of people who left clues about their existence. Archeological evidence suggests that Huber phase people lived here in 1600 AD. While the knowledge of those earliest dwellers is mostly lost, Pearion did discover a journal of a woman named Eliza Steele who traveled through the prairie in the 1840s. Her notes helped her understand which plants were here in one "original" version of this land.

We talked for a moment about the fact that the idea of reconstructing a landscape or restoring it to an "original" from some chosen earlier point in history has become even more complicated by issues of a changing climate. What lived happily on this land two hundred years ago—much less a thousand years ago—may not necessarily be the plant combination that will survive as this region warms. Pearion shrugged. "We do what we can to make it better now. And maybe some of these plants will be the ones to survive."

"Which is your favorite?" I asked.

"I look forward to seeing the charming little violet wood sorrel (*Oxalis violacea*). This plant is adorable with its tiny shamrock-like leaves and lovely violet blooms. It reminds my brain that winter is over and the prairie has started its annual march back toward an abundance of blooms. Mid-season, my eyes delight in the cauliflower-like flower heads of wild quinine (*Parthenium integrifolium*). The unusual texture and visual interest this plant's leaves and flowers offer draws me in every time."

I smiled. I knew she wouldn't be able to pick just one. "In the autumn, the striking blue of prairie gentian (*Gentiana puberulenta*) gives my heart palpitations. . . . When this little beauty is blooming, many other species are wrapping it up for the year. The colors of the prairie have begun to move away from the greens of the growing season into the yellows and russets that signal autumn, making the bright blue of the gentians all the more alluring."

Pearion pointed to a plant with enormous leaves, bigger than my laptop computer. She placed her hands on either side of one of the leaves. "Try this," she said. I knelt down and pressed the thick leaf between my palms. "It's so cool!" I exclaimed, and I meant that literally. She smiled. The leaves are cool because the roots of the prairie dock plant are about six feet deep. They pull water from far below the surface. "Rosinweed, compass plants, cup plants . . . the long roots do good work!" These plants were also important to the insects. The cup plant, for example, attracts many butterflies and bees, including *Dieunomia heteropoda*, the large black-bodied sweat bee, and fireflies. The broken hollow stems of cup plants were nest sites for leafcutter bees.

These fields were busy with bees by day and with fireflies and moths at night. I was imagining what an evening must look like, and lines from a poem by Brenda Cárdenas called "Lampyridae" floated into my mind:

> In a clearing,
> sparks burning?
> No, thousands
> of fireflies
> blinking, flickers
> of the forest's
> secrets no one
> would believe.

Hinke interrupted my reverie in to explain that prairies also sequester large amounts of carbon due to their incredible root systems. They pull carbon dioxide from the air and store it in their stems and roots. "We need to do everything possible to combat climate change at this point," she said. "And prairies are part of the answer." I looked around at the dense diversity of plants all around where we stood. It looked so healthy. Part of the reason it was healthy, I knew, though it seemed paradoxical, was because of fire. "Why do you burn?" I asked, turning to Hoffman.

He grinned. Mostly, he flew around the country putting out fires, but here he started them. I remembered watching the fire maps in California a couple of years back, when my sister's home was surrounded on three sides by voracious wildfires that were out of control. Hoffman was a member of the Hotshot team that was recruited to help: the team had particular expertise in on-the-ground strategies to control fires. Hoffman's father was the fire chief in the town where he grew up in the Upper Peninsula of Michigan. His grandfather had been a firefighter, too. The firehouse was Hoffman's hangout from a very young age, and by the time he was twelve, he was helping fight fires. The passion for fighting fire was ignited early, but his father dissuaded him from fighting house fires. He was worried about his son inhaling all of the toxins he himself had inhaled, so he encouraged Hoffman to go into the Forest Service instead. Eventually, Hoffman obtained wildland firefighting training, and in the year 2000—a bad fire season, the first of many to follow—more firefighters were required than ever before. Hoffman found his calling. Since that time, he has served as a leader in elite firefighting groups across the United States. This made him a perfect person to keep prescribed prairie burns under control.

"You burn to keep the non-native brush down," he began, "and you have to choose your burn days very carefully." Hoffman explained that every three to five years you needed a high intensity fire, hot and slow moving, to reach the cambium layer of the woody invasive brush that would outcompete the forbs and the grasses. The fire was a way of renewing the native plants. He said he wanted a day with low humidity, and this depended on how many days it had been since the last rain. He said you could not have high winds, but you could use a gentle wind if you built the fire against it. He pointed to the margins of the field where we stood, explaining how you could start at an edge, moving into the wind, or do a box burn, burning the edge first so you would know where the boundaries of the fire would be. "Of course, you don't want the fire to jump," he said, pointing to a stand of trees behind us. "Plus there is a freeway right there." Fire, in small doses, revitalized this landscape, and Hoffman made that happen safely.

We walked along a trail toward some remaining bunkers across from a fenced-off area behind which a large herd of bison wandered. Grazing animals had been introduced for a variety of reasons. Their hooves compressed the soil, and they created unique ecosystems by keeping grasses shorter. The bison helped make Midewin a refuge for grassland birds like loggerhead shrikes, bobolinks, Henslow's sparrows, short-eared owls, and northern harriers.

Nothing grew here, I thought, standing in an empty bunker seventy years after it had been filled with ammunition. From the doorway, all I could see, from me to the horizon, were tall grasses and flowers waving gently in the breeze. The word Midewin, according to the Illinois State Museum, derived from the Potawatomi word "mide," which meant "mystically powerful" and described a group of powerful healers, or "curing societies," who used their connections to animal spirits to create transformation: "Although the collective members of the Midewin had the power to cure, the individual did not." What seemed like an impossible transformation had happened here. Given the attention of a committed group of people, this land had come into a state of good health once again. The wetland experts said that the water was no longer toxic. The dragonfly populations were expanding. The creeks no longer ran red.

As we walked back toward the trailhead, Hinke and Pearion began talking about how working at Midewin had affected them. They spoke of hope. Pearion said that Midewin reminded people to stay positive. Even during the COVID-19 pandemic, this prairie offered people a place to find peace, she said. "So many losses can be healed on the trail," Hinke explained. Her father had taught her that early on. "The plants bring you back to life." Medicine, I thought.

A sudden movement on the path made me glance down. A long, elegant, pale-green praying mantis stood by my foot. I had never seen one in person. Fossils tell us that the earliest mantids lived on earth 150 million years ago. This was a resilient being. Myths about the praying mantis proliferate around the globe, connecting them with the divine, with courage, clairvoyance, and transformation. One European folktale says that when you are lost, the mantis will guide you home. I bent down and placed my hand before this regal creature, and she stepped delicately onto my palm. Mantids are voracious predators with exceptional vision. A good omen, the stories say. A soothsayer. Her triangular head tipped at an angle, and I knew she was sizing me up. Her large forelegs made her look a bit like she was wearing a cloak with deep sleeves. I knew I was in the presence of a sage. As I looked into her mysterious eyes, she seemed to say, "Together, we can survive."

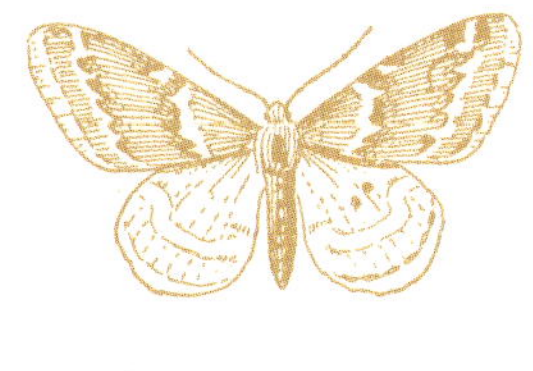

Emily Arthur

Haunted Landscapes

The act of making art, leaving a mark, keeps the balance; it holds in tension the beauty and the pain.

—NANCY MARIE MITHLO

I first learned of Emily Arthur's work when the two of us served on an art panel for an Earth Day conference in Madison, Wisconsin. I was immediately drawn into the complex images in her prints, which layered bird and insect images over maps displaying boundaries, both natural and unnatural. Ghost birds, ghost plants, maps of decimated landscapes, all bleeding into one another. All of the images called out to me like sacred texts waiting to be deciphered.

Arthur received an MFA from the Pennsylvania Academy of the Fine Arts and also studied at the Rhode Island School of Design and the Tamarind Institute at the University of New Mexico. Her pieces are included in the permanent collections of the Autry Museum of the American West, the Crocker Art Museum in Sacramento, the Minneapolis Institute of Art, the Museum of Contemporary Native Arts, and the Tweed Museum of Art as well as in collections in Japan, New Zealand, and Europe.

Arthur's energy was bright and sharp as she spoke about her artwork. A member of the Cherokee nation, Arthur's heritage deeply informs her practice. The Cherokee people from the East Coast—from places like North Carolina, Georgia, and Florida—were forcibly relocated to the territories of Oklahoma. Arthur's work is often about dislocations, exile, and about what has disappeared, heritage and humans and non-humans, from her own people to specific species of plant, insect, and bird. Her work seems to be trying to solve some riddles: How do you grieve an invisible loss? How do you make that invisible loss visible? How do we, in this contemporary moment, engage deeply and honestly with our haunted landscapes? How best can we mourn and honor what's gone, and then protect what's left?

The power of printmaking, in part, is its palimpsestic aspect. Layers of ink and image can obscure other layers, the way erosion or development can obscure earlier ecosystems and communities in our landscapes. What is written on the earth by human activity can be covered over, removed, eroded, and erased, making it hard to remember, hard to envision, hard to restore. But printmaking also builds something new out of all the layers of ink. Arthur's haunting prints invite us to remember cultures and nonhuman beings who have been displaced or pushed to extinction.

In a series Arthur created during a residency with the Moore Laboratory of Zoology in Los Angeles, in collaboration with John McCormack and James Maley, many prints have images of dead birds lined up as they might be in a drawer of a zoological collection. When she was working in the lab, Arthur learned how the protection of the California gnatcatcher, a species listed as endangered, was inhibiting developers and how scientists were being encouraged to manipulate their data to suggest that the bird was no longer in danger. A change in government policy might allow development to move forward and also lead to the eradication of certain species. As Arthur explained in an interview, "This is my first experience with scientists who share with me their data collection and expose how scientific data can be manipulated to control the outcome in government policy."

The fact that, historically, government policy can lack justice or can have terrible consequences is obvious to anyone with an awareness of Cherokee history. In the late 1700s and early 1800s, a large portion of the Cherokee Nation lived in northwestern Georgia. When Andrew Jackson occupied the office of president, the state of Georgia began passing many laws prohibiting the Cherokee from owning property or mining, and in 1835, the Treaty of New Echota—which was signed by a small faction of the Cherokee—forced the removal of all Cherokee people east of the Mississippi River. This treaty was not signed by the chief, nor was it approved by the Cherokee National Council. The forced removal, known as the Trail of Tears, led to the death of approximately four thousand people.

Arthur's keen awareness of erasure, born in part out of the stories of her ancestry and in part out of her passion for the natural world, fuels her creative projects. When she became aware of a Florida marsh that was being dredged and replaced by a shopping mall, she took large pieces of paper to the site. She had the construction vehicles drive over them, leaving tire tracks on the paper, on which she then superimposed images of the flora and fauna that were being destroyed. Arthur called the series *Once Wetlands*.

Several of the images featured in this gallery focus on the Palos Verdes blue butterfly (*Glaucopsyche lygdamus palosverdesensis*), which, like the California gnatcatcher, depended upon a plant called coastal sage to survive. Entomologist Richard Arnold observed that the Palos Verdes blue butterfly was "endemic to the Palos Verdes Peninsula near Los Angeles, California, USA, where the butterfly or its sole larval foodplant, *Astragalus trichopodus* var. *lonchus* (Fabaceae), formerly occurred in eighteen sites in association with a coastal sage scrub plant community." The disappearance of coastal sage, owing to the

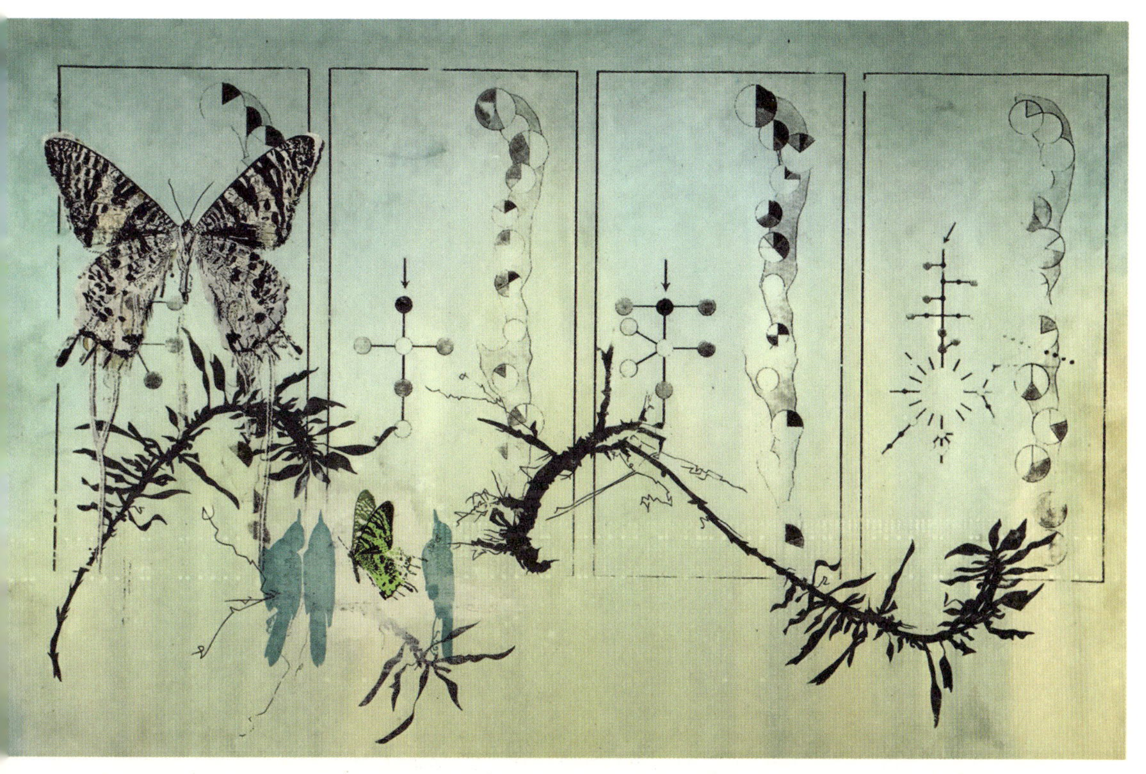

Emily Arthur, *Tiger Moth on Sea Green*, 2017. Unique screenprint on dyed Rives paper, 22 × 30 in. Image courtesy of the artist.

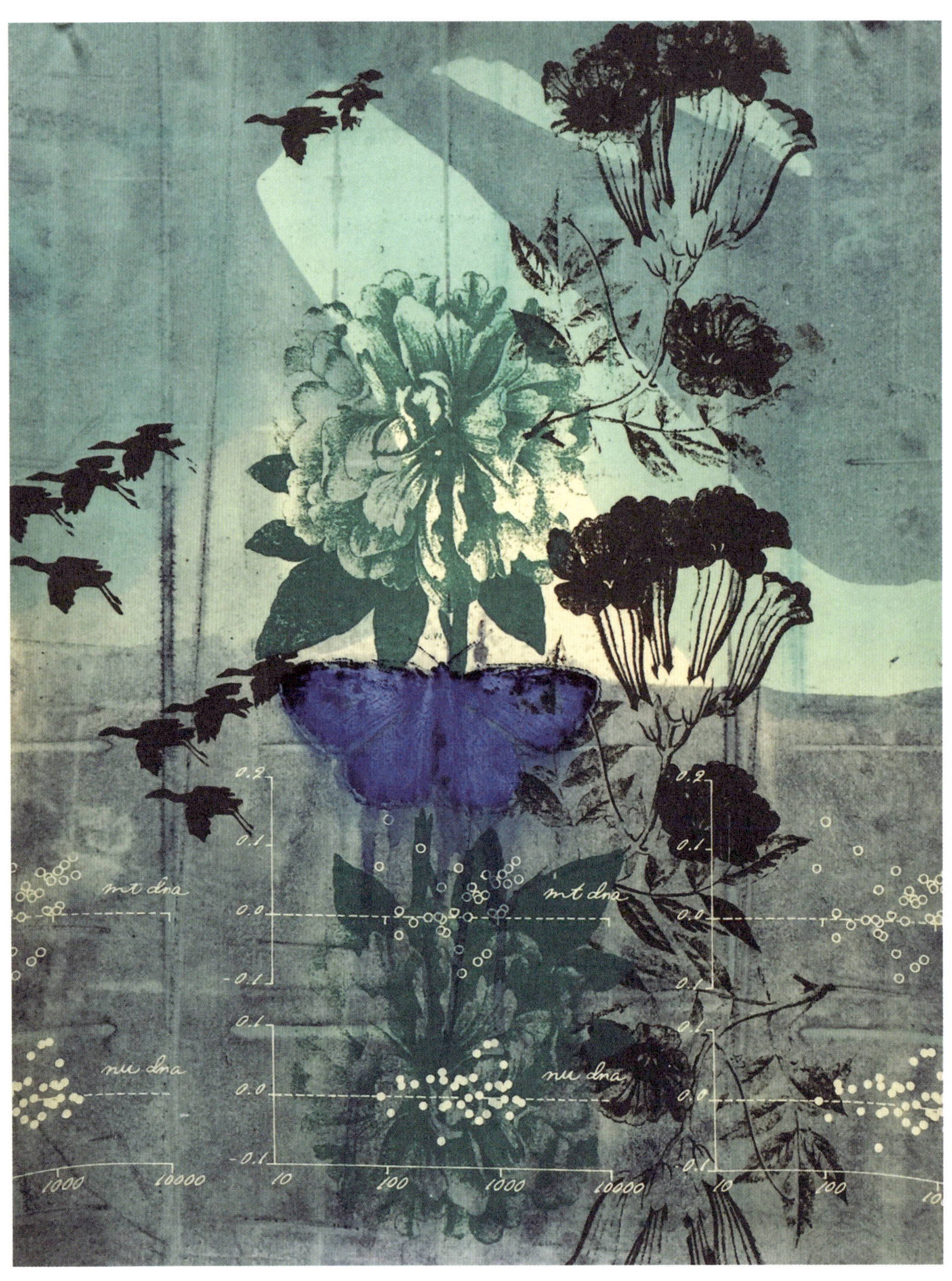

Emily Arthur, *Palos Verde Blue with Peony*, 2017. Unique screenprint, 20 × 16 in. Image courtesy of the artist.

development of recreational and housing areas, had made it nearly impossible for the Palos Verdes blue butterfly to thrive. In the 1980s, its population had pitched down by 90 percent. The butterfly was considered one of the world's rarest and was in danger of extinction. In 2020, however, one thousand of these butterflies, which were bred in captivity, were released onto the Palos Verdes Peninsula near Los Angeles in habitats restored by the Palos Verdes Peninsula Land Conservancy. There is hope for healing if we become aware of these complex relationships and try to protect them. Arthur's work helps make us aware.

I am reminded of a passage in Lydia Millet's novel *How the Dead Dream*. Her fiction often highlights problems of extinction. Millet writes poignantly about what humans will miss as extinction accelerates:

> And yet a particular way of existence was gone, a whole volume in the library of being. Others were sure to fall afterward—a long fly with iridescent wings that lived only in the nest of this single rat, say; a parasite that lived under the wing of the fly; a flowering plant whose roots were nourished by the larval phase of the parasite; a bat that pollinated the plant . . . it was time that would show the loss, only time that would show how the world had been stripped of its mysteries, stripped by the hundreds and thousands and millions.

Emily Arthur awakens her viewer to beauty and pain. Each time I revisit her work, I find myself asking: *What, then, must we do?*

My heart does know one thing: we cannot stop listening to the stories of our elders and the stories of all beings, and we cannot stop working for justice for all.

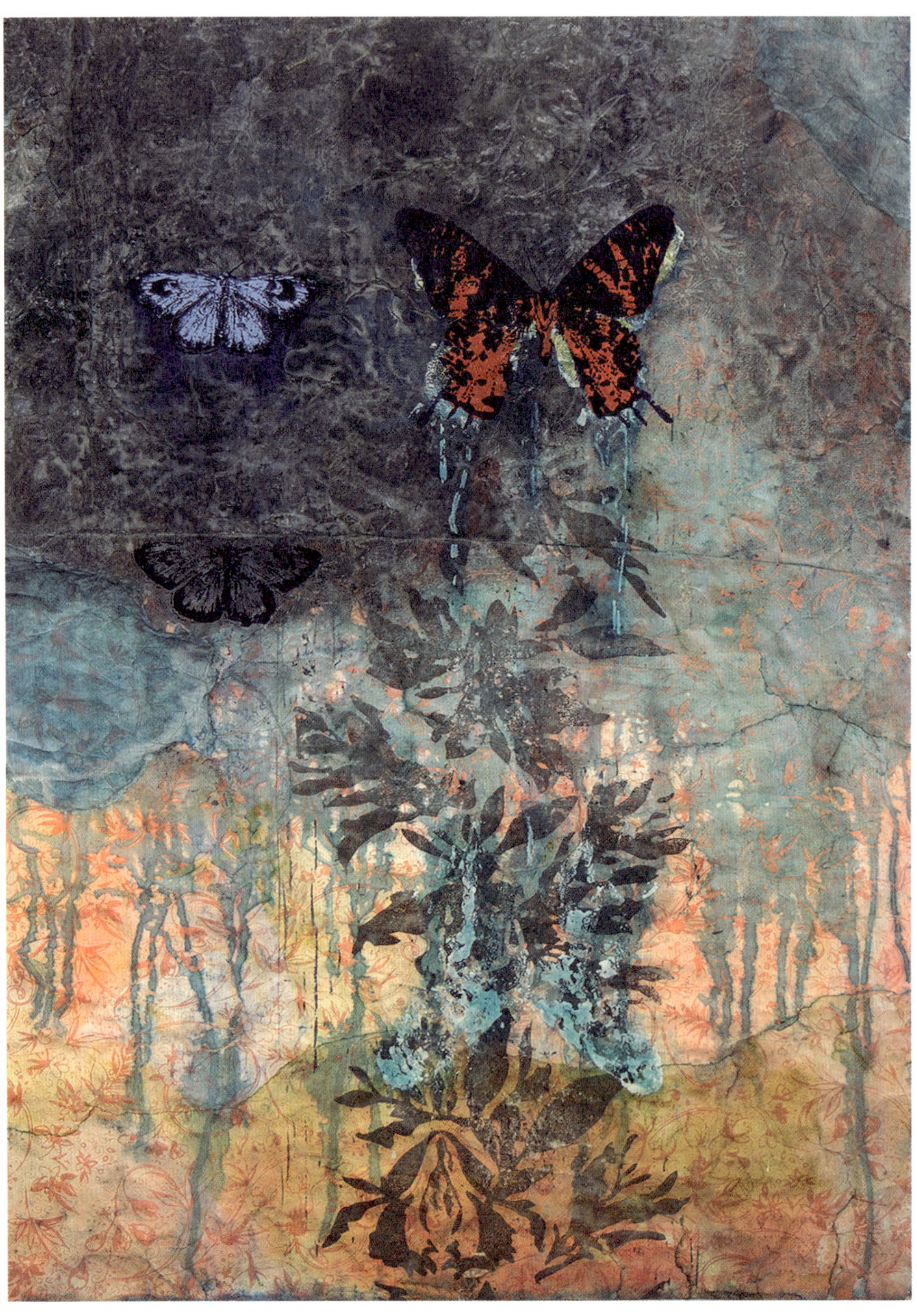

Emily Arthur, *Palos Verde Blue with Tiger Moth*, 2017. Unique screenprint with glazed pigment on hinged sheets of dyed Rives paper, 40 × 30 in. Image courtesy of the artist.

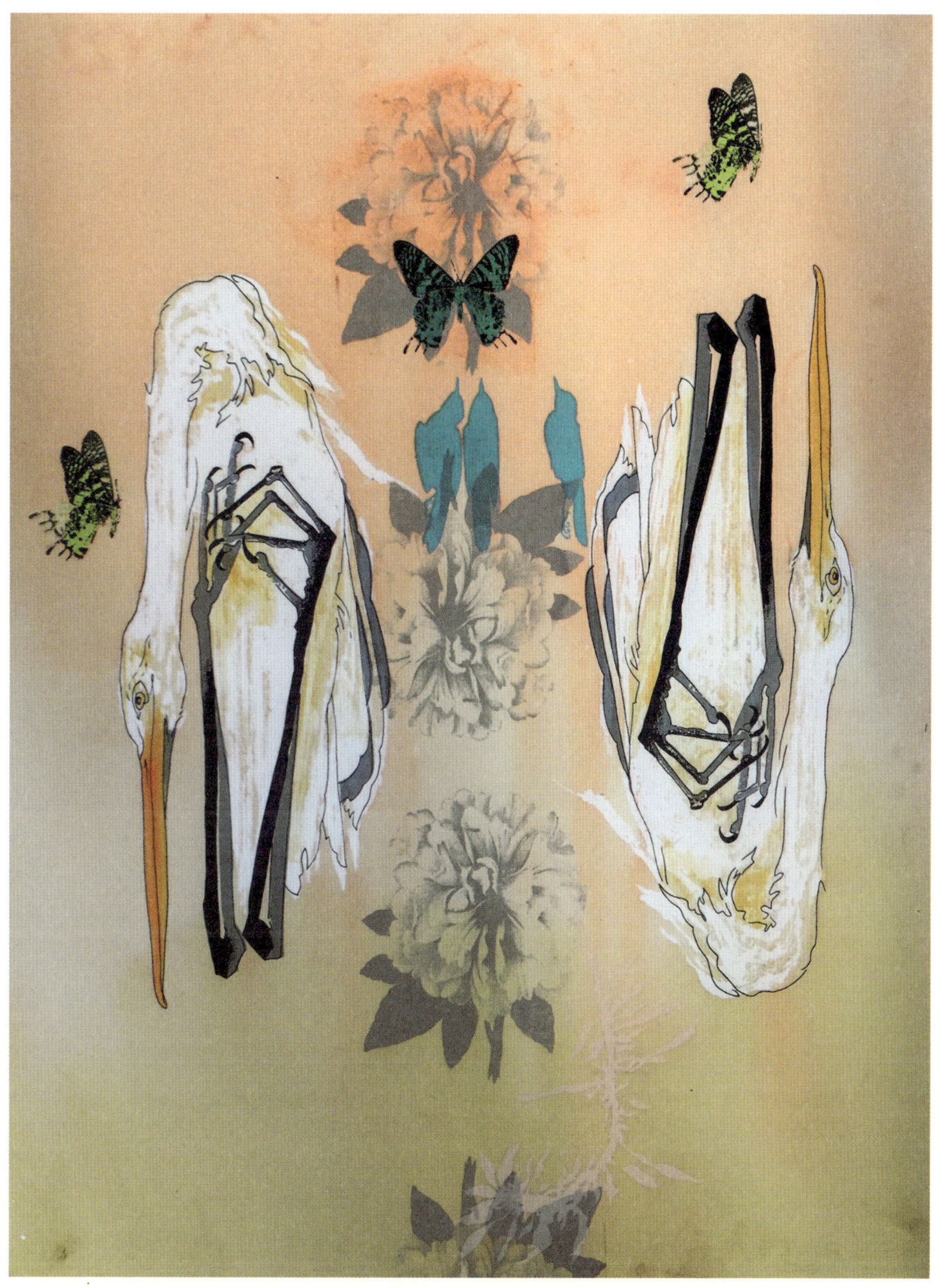

Emily Arthur, *White Egret with Peony*, 2017. Unique screenprint on dyed Rives BFK paper, 30 × 22 in. Image courtesy of the artist.

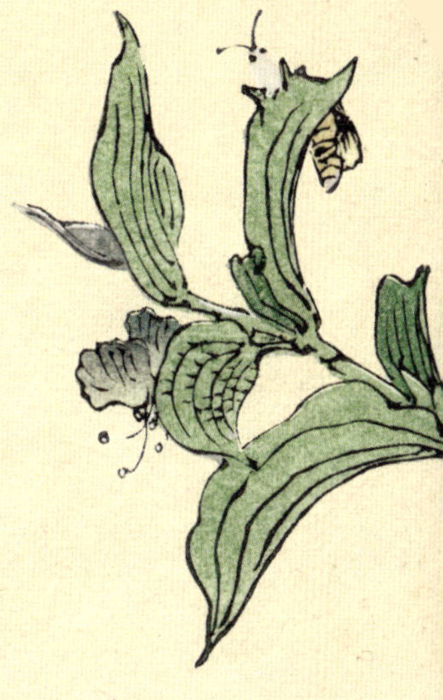

The Bumblebee and the Cranberry

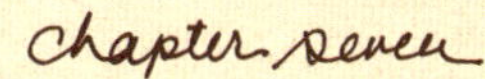

> Bumble bee, bumble bee, don't be
> gone so long.
> You're my bumble bee, and you're
> needed here at home.
>
> —MEMPHIS MINNIE,
> "BUMBLE BEE BLUES" (1930)

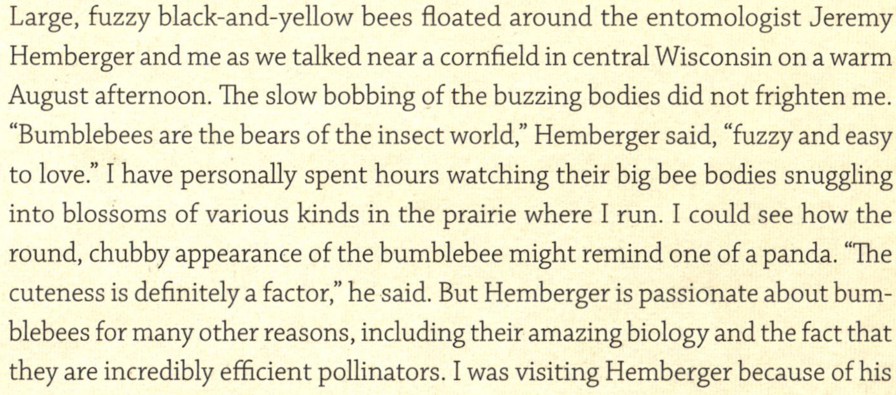

Large, fuzzy black-and-yellow bees floated around the entomologist Jeremy Hemberger and me as we talked near a cornfield in central Wisconsin on a warm August afternoon. The slow bobbing of the buzzing bodies did not frighten me. "Bumblebees are the bears of the insect world," Hemberger said, "fuzzy and easy to love." I have personally spent hours watching their big bee bodies snuggling into blossoms of various kinds in the prairie where I run. I could see how the round, chubby appearance of the bumblebee might remind one of a panda. "The cuteness is definitely a factor," he said. But Hemberger is passionate about bumblebees for many other reasons, including their amazing biology and the fact that they are incredibly efficient pollinators. I was visiting Hemberger because of his work with *Bombus impatiens*, a bumblebee native to Wisconsin, and in particular, its relationship to cranberries.

Thousands of years ago, the Laurentide Ice Sheet, the last large glacier to mold the soils and rocks of Wisconsin, smoothed the land where we stood, making it a perfect place for growing crops that require a lot of land. To the west of us, drumlins and moraines rise out of the plains with creeks and rivers veining between them, leading eventually to the unglaciated area, lovingly called "The Driftless." The Ho-Chunk claimed this land as home for centuries before European settlers invaded. Burial mounds built by Indigenous communities to honor the native animals of Wisconsin remain as evidence of their connection to this place, even hundreds of years later. The land changed drastically over time under the watch of the settlers. Cropland and farms exploded as technology changed what agriculture looked like—most dramatically with the onset of our current industrial model of farming. Today what surrounds us are corn and soy, for the most part. Not more than an hour's drive from Hemberger's experimental site, Aldo Leopold famously took some old farmland that he felt had been abused and restored it, bringing back prairies and oak trees. Out of that project and his thinking in Wisconsin came his seminal work, *A Sand County Almanac*, published in 1947. "We abuse land because we regard it as a commodity belonging to us. When we see land as a community to which we belong, we may begin to use it with love and respect," Leopold wrote.

I look out over the cornfields stretching to the horizon. These rows of green are deceptive, I know. They seem fertile; the green makes it seem like a natural space. But, in fact, the fields represent commodity thinking, not so good for the community of humans and nonhumans. Insects and birds need more than chemically supported corn to survive, but monocrops are what were developed here, in this part of Wisconsin. Farther north, crops change as the landscape changes. In northern Wisconsin, wetlands and lakes dapple the land, and that is where cranberries flourish. Unlike corn, cranberries need pollination by insects. The *Bombus impatiens* drowsily drifting around us were a part of an experiment Hemberger had been working on since early in the year in his lab at University of Wisconsin–Madison. *Impatiens* have been used by cranberry growers in Wisconsin more and more in the last few years as an alternative to honeybees for pollination. I wanted to know more about why and what, specifically, his work suggested. How did the bumblebee affect cranberry production?

Hemberger is an entomologist but wasn't always a bug person. He started in genetic molecular engineering and worked for a time on brown trout biogenetics and West Nile virus in mosquitoes. Eventually he realized he wanted to work on ecology and conservation biology. As awareness of pollinator decline grew, he decided he would focus on wild bees. His particular passion was the bumblebee.

Bumblebees live in communities, like honeybees, the bees I was most familiar with. They nest in the ground and raise baby bees and even make honey, though much less than a honeybee does. The bumblebee queen begins her home-making in the spring on her own, building a nest and laying eggs until she has a strong family for the summer season. Bumblebee nests have some very stark differences from those of the honeybee. One difference is that honey is stored in little wax pots in a bumblebee colony. A honeybee builds a wall of comb made up of hexagonal cells, which are filled with honey and capped. The honeybee hive's design and the size of the colonies have made stealing honey from honeybees very easy for humans. Bumblebees nest in the ground in smaller numbers, and so their honey production is lower. These qualities made them less interesting to humans who wanted honey. Bumblebees also spin silk cups for their babies. Bumblebees are especially interesting to Hemberger because they can regulate their temperature. A bee can pump fluid into her belly and either heat or cool herself when the weather is especially cold or hot.

The ability to regulate body temperature is important, especially in a climate that has as much temperature variation as Wisconsin. But even this trait is not enough to protect bumblebees from the challenges that will come with climate change. In the spring of 2020, a study in *Science* magazine showed that bumblebees are in fact in serious decline, not only because of pesticides (which had been clearly shown by the British entomologist Dave Goulson) but also because of the earth's rising temperatures. Not only does the insect herself suffer, but habitats that would support her could also dwindle dramatically, leaving a dearth of food.

When I arrived, Hemberger was standing near his makeshift field lab in the back of his pickup truck. Styrofoam cases, vials, and various tools were arranged on the truck's tailgate. He hunched over three vials in his white coveralls, looking like a cartoon version of a mad scientist. When I looked closely, I could see that there was one bumblebee in each vial; they were flying erratically in apparent attempts to escape. Hemberger wanted to photograph and measure them, so he would have to slow them down. A quick reduction in temperature would do this. He opened up a cooler full of ice and plunged each vial down inside for a quick chill. After a minute or two, he pulled them back out, and sure enough, the bees had slowed to the point of immobility and appeared to be tiny insect statues.

I thought about a video I had seen of a bee sleeping in a flower, as Hemberger had told me bumblebees sometimes did, her furry body nestling between petals and stamens to rest while her portion of the earth tipped away from the sun. Did bees dream? I wondered. Maybe of great fields of color? Were their nightmares of planes spraying their homes with chemicals? These bees weren't sleeping, though. They were frozen.

He assured me that this rapid cooling would not harm them at all and that they would quickly warm up with the air temperature. He placed each of them on a plate next to a tiny battery, for scale, and took a quick picture. At this site he has gathered information each week: hive size, number of brood, honey production.

While people often think of honeybees as the most crucial pollinators to save, many entomologists would argue that this kind of thinking has led us to forget that native pollinators, especially wild bees, are essential to the health of a landscape and can also serve as formidable pollinators for crops. The industrial agricultural model utilizes honeybees for pollination, in part because they are so easy to control. They live in much larger colonies, for one thing. Honeybee colonies are made up of approximately fifty thousand bees, while bumblebee nests only have two hundred to three hundred worker bees. Honeybees have also adapted to living as livestock in Langstroth hives. The Langstroth hive is made of wooden boxes filled with rectangular frames, which the bees fill with comb. This is unnatural to the bees themselves, but the design allows a beekeeper to get in and out of a hive easily to collect honey and check on the bees. The boxes are also easily stacked and shipped. Every year millions of beehives are stacked on the backs of semitrailer trucks and driven from huge monocrop to huge monocrop. Cranberries are a part of that system.

One destination is the almond groves of California, which are pollinated by millions of honeybees. When the trees are covered with pink blossoms, and the flowers are covered with pollen and ready for pollination, honeybees are brought in to work the fields. Beekeepers from around the country stack their hives at the edges of the groves, and the honeybees begin collecting nectar and pollen to bring back to their individual hives. The bees work tirelessly until all the almonds are pollinated and they have exhausted the pollen source. But then they have nothing to eat. Until very recently, the miles and miles of trees had nothing growing between the rows. The lack of other flowers meant that the honeybees had nothing else to eat after the almond pollination was completed. (Beekeepers often supplement the bees' diet by feeding them high fructose corn syrup.) Additionally, many of the trees have pesticide residue, which is also consumed by the bees. A honeybee does not thrive in these conditions.

The use of pesticides is part of our culture. We love them. We use them in our homes. We feel comfortable eating them on our food. We pour them on our lawns. In 2020, we spent $105 billion on our lawns; an estimated $40 billion of that goes to fertilizer and pesticides. Homeowners spray their lawns with up to ten times the amount of pesticide per acre than farmers use on their fields. We seem to have forgotten the warnings of Rachel Carson's *Silent Spring*, which memorably exposed readers to the devastating effects of DDT. She loved the earth. She found wonder and delight in its myriad creatures. She loved tide pools and birds. She offered us wonder and warning. I remember reading her work and thinking, How could we still be making the same mistakes over fifty years later?

When my daughter was still small, maybe two years old, I remember once waking at dawn to the sound of an airplane barreling over our house. The roar grew intense and loud and then began to fade as it was flying away from us. When I heard the plane turn around, with the sound of menacing engines growing louder and louder, I grabbed my daughter and ran downstairs. The memory of 9/11 was still fresh in my mind, and so it occurred to me that the pilot might be intending to terrorize someone or maybe crash intentionally. Or perhaps it was a local pilot who had lost their mind? The plane roared over and away again. Panic guided me down the stairs, and for some reason, I went out the back door onto the deck so I could see better. The trees overhead obscured my view of the plane, but I could hear it turning around again. Why would I have stayed there? Maybe because I thought I could get a good look at the plane and identify it later if necessary? Maybe I thought I should not be inside a structure if it was going to crash on us? I am not sure what made me stay there; adrenaline pulsed through my body and guided my brain. But when the aircraft came back, I could see it clearly.

A plane the color of a yellow cab zoomed over the tips of the pines and the maple growing in my yard, and as the belly of the plane sheared the treetops, I saw, too late, that it was spraying. My body reacted too slowly to avoid being coated in a thin mist. My baby and I were covered in a pesticide. Horrified, I ran inside. Moments later we were in the shower, my body shaking with rage and fear.

Two weeks later, one of my daughter's lymph glands in her neck had grown to the size of a mango. The chemical they sprayed carried warnings about those with asthma. Exposure could be problematic. My daughter had asthma. The doctors said that there was not enough evidence to prove that the chemical had caused the lymphatic response, but I knew inside that's what had happened. Luckily her response had not been worse.

The experience of being sprayed like that, being unable to protect myself or my child, and then watching her suffer afterward gave me insight into what bees, frogs, and birds experience all the time. There is no escape. Farmworkers are often unwittingly exposed to pesticides that can cause Parkinson's disease, cancer, and many other problems.

The effects of neonicotinoid pesticides, which are neurotoxins, have been examined quite extensively in honeybees, and it is well known that the insects are suffering with problems of navigation and overall health. Although there has been little done to reduce the use of these pesticides in the United States, the science is quite clear. These chemicals are bad for honeybees. What is less well known is their effects on bumblebees and other insects.

Hemberger shared a new study out of Harvard that revealed that neonicotinoids greatly changed the social behavior of a bumblebee. As Hemberger explained to me, normally, one solitary female bumblebee would begin a colony. She would weave silk cases for her eggs and incubate them herself. Once the colony was larger, never more than a couple hundred bees (unlike a honeybee colony, which can grow into the tens of thousands), the offspring would start to participate in hive maintenance. The effects of neonicotinoids on bumblebees, though, made them antisocial. Affected bees would remain at a distance from hive activity, rendering them useless to the longevity of the hive. While much of the focus of the media is on the decline of honeybees, other bees, including bumblebees, are also in serious trouble.

The difficult issues surrounding the use of the honeybee for large-scale pollination projects and the fact that they are obviously suffering have led some people to question the system itself. How effective is it to "save the honeybee" when the pollination system we have created for them is, in great part, responsible for their demise? One part of the solution is to plant a more diverse habitat around our existing crops so that the honeybees have more to eat and are better nourished. Another is to grow smaller-scale organic crops. But another idea that many entomologists are getting behind is to lean less on the honeybee and begin to rely more on wild bees. This is what Hemberger was interested in.

The three chilled *Bombus impatiens* had warmed up and had taken to the air, flying in a drunken waltz, only slightly irked, it seemed, by the inconvenience of having to slow down and be measured. They seemed healthy, Hemberger thought. We walked over to the side of the field and stood over a wooden box on the ground. He had placed a bumblebee nest inside the box. He explained that one way to monitor the foraging habits of a bumblebee is to attach a tiny radio transmitter to her back. The wooden hive box has a device attached to the opening that recognizes each bee as she returns from foraging. His plan here is to establish whether planting a field of flowers next to the corn would help the bees' survival.

When I visited him at Claudio Gratton's lab in Madison, Wisconsin, earlier in the year, Hemberger had shown me samples of the many kinds of *Bombus* that live in Wisconsin. The lab is filled with state-of-the-art equipment for examining insects and metal racks covered with boxes filled with collected samples. Hemberger pulled down a few of his collections. He carefully lifted the lid of one box after another. Inside were rows of fuzzy bumblebees of differing sizes, floating motionless above the floor of the box on pins. *Bombus ashtoni*, with a thin white stripe; *Bombus ternarius*, with a wide orange band; *Bombus affinis*, with a tiny orange spot on her back (commonly known as the rusty patch bumblebee, which has gained notoriety for being endangered); and *Bombus impatiens*, sporting an almost completely yellow abdomen. Each had been washed, dipped in ethanol for preservation, and blown dry. As in the collection I saw in the Essig lab at Berkeley with Peter Oboyski, each insect had a label that identified the type of bumblebee as well as where and when she had been collected. Some boxes had as many as 120 bees inside. The bees were not arranged with aesthetics as the priority, as these bees were not for public display. But Hemberger also had live bees.

He led me down the hall from the lab and opened a door to reveal a glowing red room with deep glass-covered cases along one long wall. The cases contained several bumblebee nests. The bees could not see us due to the color spectrum their vision allowed, he explained, but we could see them. Big fat bees crawled calmly around their nests inside a large glass case. The experiment Hemberger had been conducting in the lab involved looking at bee health and their diets over time, not dissimilar to the fieldwork. How would the bees do if their diets surged now and then? What if there was a dearth of flowers? The real goal of all of this was to see how important bumblebees were—or could potentially be—to a farmer who required pollination.

While many farmers hired honeybee keepers to bring bees to their crops, other farmers had recently thought about returning to native bee support. I wanted to talk to a farmer who had actually used bumblebees for pollination. Hemberger suggested that I go meet Corey Searles, a fourth-generation cranberry grower who lived in northern Wisconsin. The sun was low in the sky when I drove away from Hemberger's truck, my head still buzzing from all I had learned and the questions I still had.

That October, I stood clad in rubber hip boots thigh-deep in the very chilly waters of a Wisconsin cranberry bog, surrounded by an acre-wide floating blanket of bright red and pink berries. Corey Searles had handed me the tall boots and a wide rake when I arrived and asked me if I would wade in and help. The gray sky and cold wind made the idea of being in the water daunting, to say the least, but no one else was complaining. Corey shouted introductions to the team across the water. At work that day were Corey's brother, his aunt Becky, his dad, and Chuck, a good friend of the family. When Hemberger had told me that this was a family operation, I hadn't actually expected it to be so true. Soon I learned that the late Clarence Searles, Corey's grandfather, had pioneered modern cranberry farming in this part of Wisconsin.

Before Clarence Searles's time, one could find native Wisconsin cranberries growing in the bogs, but none were cultivated as commercial crops. The Ojibwe called this plant *mashkiigimin*, or swamp berry. The cranberry had long been used as a medicinal superfood for communities like the Ojibwa or Anishinaabe, and in northern Wisconsin and Minnesota, they are still very much part of the diet. In a report created for the Milwaukee Public Museum called "Ethnobotany of the Ojibwe Indians," Huron Smith, who had gathered information from elders during three trips to Ojibwe reservations between 1923 and 1924, wrote that "according to the Ojibwe, every plant is medicine. . . . Consequently, in the field we gathered every tree, shrub, perennial or annual, herb or grass we encountered. All of these being medicinal plants were thus sacred to them and must be secured with the proper mide ceremony." In his report, Smith lists many specific plants and their traditional uses. Prickly ash, or *gawa kumic*, helped with bronchitis and respiratory problems. The shining willow, or *muckigo bamic* ("swamp tree"), stemmed bleeding and could be used as a poultice. Black snakeroot, or *masan*, cured a rattlesnake bite, but it was used by some to cure fever, sore throat, and "nervous affections." Cranberries were made into "tea for a person who is slightly ill with nausea." Today, the cranberry is known by many nutrition enthusiasts to be high in antioxidants. Cranberry sales soar during the American holiday of Thanksgiving, a holiday fraught with issues of colonialism. For many, it is a reminder of the injustices experienced by the First Nations. Corey Searles would prefer to have people eating cranberries year-round.

Corey said his grandfather Clarence hiked through the woods and cut sections from vines that looked good to him. He rooted them and tied colored threads around their stems, creating a system for identifying which ones brought one berry or two and which berries seemed best. Through this process, he found vines that produced the best-quality fruit. The result was a vine named the Searles Vine, which many growers in this area began to use. The fact that these cranberries had existed here for thousands of years made a native insect like *Bombus impatiens* an obvious partner for pollination.

The cranberry "bog" we stood in was not a true bog. While there are many true bogs in northern Wisconsin, these flooded fields were human-made depressions in the earth equipped with giant pumping systems, which circulated water on and off the vines that grew there. I had visited a true bog with the botanist Susan Knight a few years before and had seen wild cranberries growing there, along with many unusual carnivorous plants. Layers of peat under our feet had created a mat of decaying plant matter that actually gently rode up and down when you stomped on it. With the Searles family, though, I was standing in thigh-deep water on solid ground planted solely in cranberry vines.

The object of the workday was to harvest several tons of ripened cranberries and load them into the huge semi parked at the side of the field. My job was to rake the floating red berries across the water toward a two-foot-by-two-foot basket, which had a suction mechanism inside it. It was like a large monster sucking all the cranberries out of the water and into a giant tube.

A machine with long metal fingers loosened the berries from the vines in the flooded beds, and these floated to the top. We pushed the berries to the tube, which sucked the berries into a sorting machine. These cranberries would be shipped to Ocean Spray and Mariani.

Later, in the kitchen of the Searles home, Corey told me that when he had learned about the honeybee crisis called Colony Collapse Disorder, he was worried, so he hoped he could count on other bees for pollination support. Some of these pollinators are even better for cranberries than honeybees. The entomologist Marla Spivak, from the University of Minnesota, explains in an extension pamphlet for cranberry farmers, "Honey bees are the most effective pollinators of cranberries, but bumblebees are the most efficient." How do we bring back our bumblebees and keep them around? Jeremy Hemberger's bumblebee research suggests that bumblebees need more food throughout the year and that "increasing total food abundance will have the greatest, positive impact on colony fitness."

Working with the University of Wisconsin extension service, Corey redesigned some areas of his land so that native plants could thrive; these plants would support other pollinators. His home is surrounded by a patchwork of woods and native plants and cranberry bogs, providing homes for many animals and insects besides the pollinators.

Before I left, Corey had me try a special treat: a hand-dipped caramel-covered cranberry, a perfect balance of sweet and tart. As I left the Searle family home, I wondered about other farmers. What else could be done?

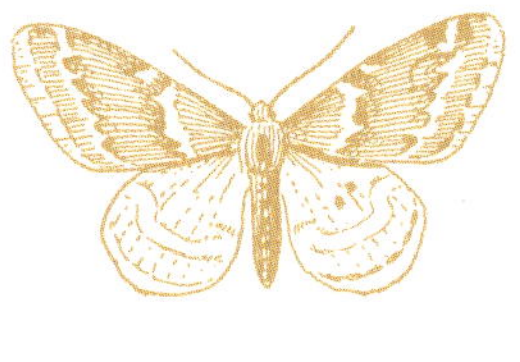

Liz Anna Kozik

Illustrating Interconnections

Attention becomes intention, which coalesces itself to action.
—ROBIN WALL KIMMERER

Children huddled eagerly around Liz Anna Kozik as she painted highly detailed flowers and all sorts of bees in her studio at the Discovery Center at the University of Wisconsin–Madison during a science festival. The walls and windows had been plastered with sketches and fully rendered illustrations of prairie plants. Even the arms of the artist herself were covered with tattoos of her favorite prairie plants. Her mission was to share everything she loved about plants and to educate people about why they are valuable. "People like pictures," she said, so drawing and painting are her tools for education.

Born in Illinois, Kozik grew up on prairies, but not until she was going to school at the Rhode Island School of Design did she really begin to miss and value them. Eventually she landed a job in Chicago doing product design, but knew she needed to work on things closer to her heart rather than continue as an "ugly Christmas sweater spreader." An opportunity to study in Wisconsin allowed her to bring her gifts into alignment with her values. Soon she was making graphic storyboards to tell stories about the importance of prairies, their plants and insects, and the people like Aldo Leopold who championed their restoration. Her master's thesis, "Stories in the Land," beautifully narrated the story of Curtis Prairie.

On a warm day in early spring, we met near Curtis Prairie in Madison. Kozik chatted with me absent-mindedly as we walked through a wooded area, interrupting her sentences to greet the many spring ephemerals we were passing. "Oh, I know who this is!" she would exclaim, reaching down to delicately touch a white trout lily or Dutchman's breeches, greeting them under her breath. Her conversations with the plants brought to mind the words of one of our mutual mentors, the Potawatomi botanist Robin Wall Kimmerer: "Recognition of the personhood of other beings asks that we relinquish our perceived role as masters of the universe and celebrate our essential role as an equal member in the democracy of all species." Like Kimmerer, Kozik reminds me of the importance of naming things and says we must remember that the knowledge of these small plant beings is ancient. We modern folk have simply forgotten. She talked a bit about her desire to teach people but then was distracted by some lovely trillium and fringe toothwort. She knows this landscape like a close friend and has painted nearly every flower we meet.

Liz Anna Kozik's truest loves, however, are prairie plants, and so she is keenly aware of their collaborators, native pollinators. Through painting and adoration she has come to know volumes about the prairie. "Knowledge is of no use if it's not shared," she said. Sharing her artwork about the need to save prairies is her activism. At the time of this writing, Kozik had been busy raising awareness with her art on social media about how the Bell Bowl Prairie in Minnesota was at risk of being destroyed by a neighboring airport. This prairie is one of the few habitats left of the rusty patch bumblebee. For the moment, in large thanks to Kozik, that habitat remains.

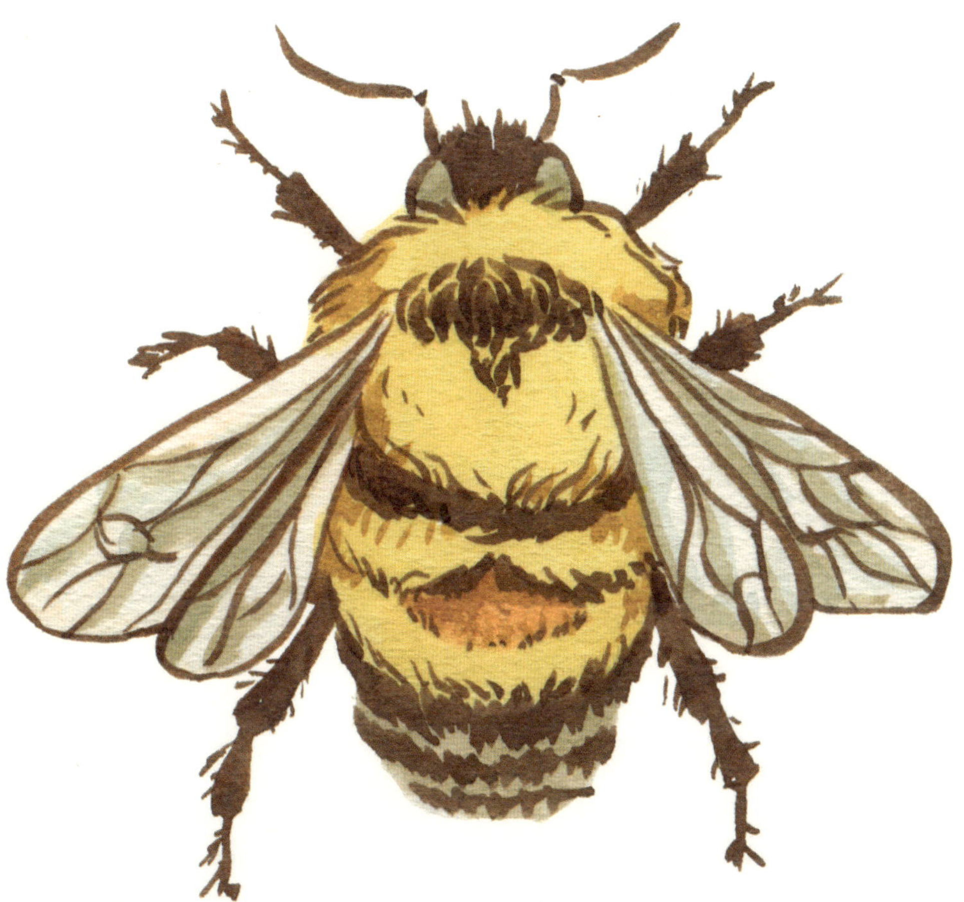

Liz Anna Kozik, *Bombus affinis*, 2018. Pen and ink.

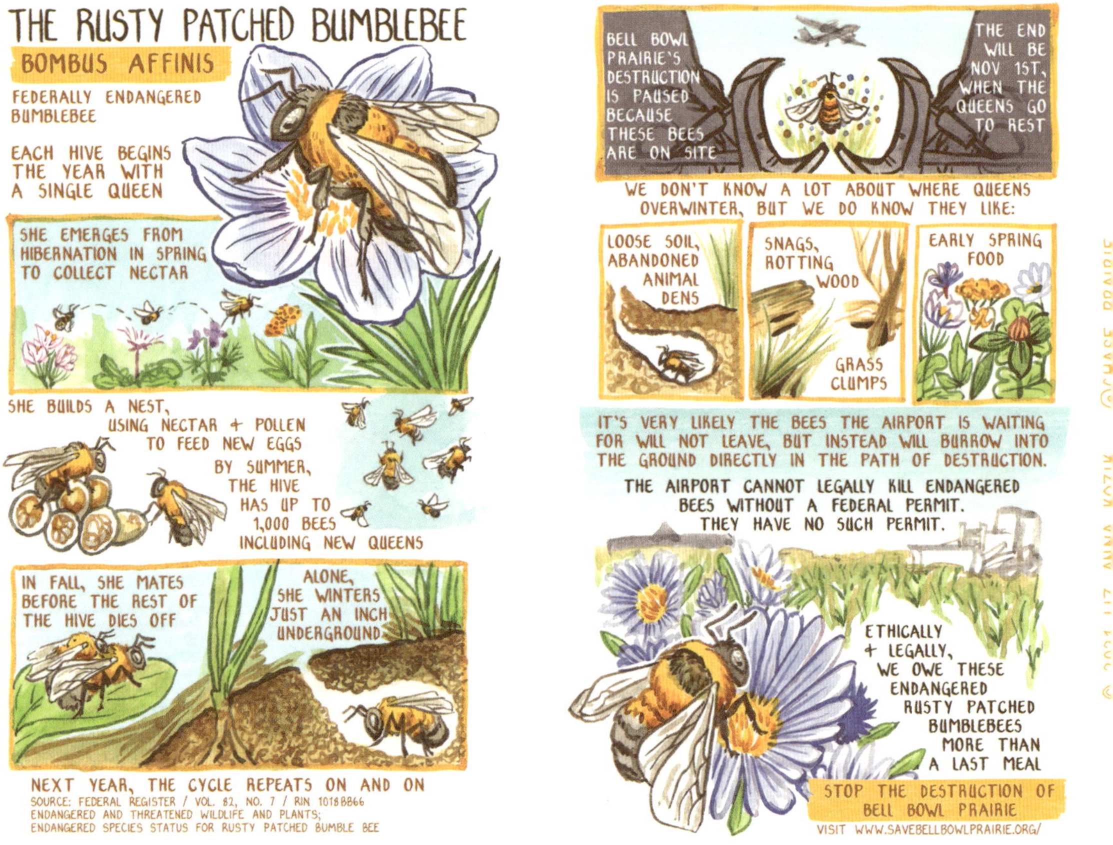

THE RUSTY PATCHED BUMBLEBEE

BOMBUS AFFINIS

FEDERALLY ENDANGERED BUMBLEBEE

EACH HIVE BEGINS THE YEAR WITH A SINGLE QUEEN

SHE EMERGES FROM HIBERNATION IN SPRING TO COLLECT NECTAR

SHE BUILDS A NEST, USING NECTAR + POLLEN TO FEED NEW EGGS

BY SUMMER, THE HIVE HAS UP TO 1,000 BEES INCLUDING NEW QUEENS

IN FALL, SHE MATES BEFORE THE REST OF THE HIVE DIES OFF

ALONE, SHE WINTERS JUST AN INCH UNDERGROUND

NEXT YEAR, THE CYCLE REPEATS ON AND ON

SOURCE: FEDERAL REGISTER / VOL. 82, NO. 7 / RIN 1018 BB66 ENDANGERED AND THREATENED WILDLIFE AND PLANTS; ENDANGERED SPECIES STATUS FOR RUSTY PATCHED BUMBLE BEE

BELL BOWL PRAIRIE'S DESTRUCTION IS PAUSED BECAUSE THESE BEES ARE ON SITE

THE END WILL BE NOV 1ST, WHEN THE QUEENS GO TO REST

WE DON'T KNOW A LOT ABOUT WHERE QUEENS OVERWINTER, BUT WE DO KNOW THEY LIKE:

LOOSE SOIL, ABANDONED ANIMAL DENS

SNAGS, ROTTING WOOD

GRASS CLUMPS

EARLY SPRING FOOD

IT'S VERY LIKELY THE BEES THE AIRPORT IS WAITING FOR WILL NOT LEAVE, BUT INSTEAD WILL BURROW INTO THE GROUND DIRECTLY IN THE PATH OF DESTRUCTION.

THE AIRPORT CANNOT LEGALLY KILL ENDANGERED BEES WITHOUT A FEDERAL PERMIT. THEY HAVE NO SUCH PERMIT.

ETHICALLY + LEGALLY, WE OWE THESE ENDANGERED RUSTY PATCHED BUMBLEBEES MORE THAN A LAST MEAL

STOP THE DESTRUCTION OF BELL BOWL PRAIRIE

VISIT WWW.SAVEBELLBOWLPRAIRIE.ORG/

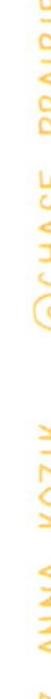

Liz Anna Kozik, *Bell Bowl Bees*, 2021. Print, pen and ink.

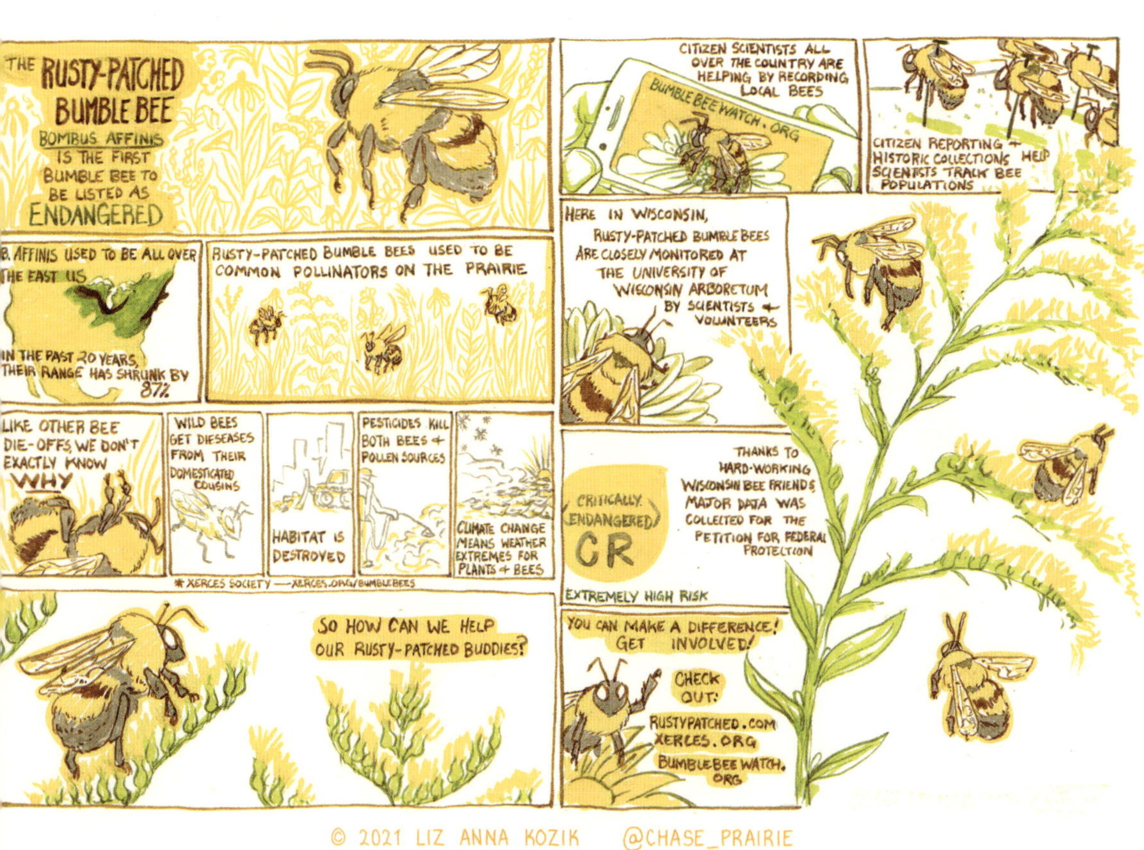

Liz Anna Kozik, *The Rusty-Patched Bumblebee*, 2017. Pen and ink.

NATIVE BEES OF NORTH AMERICA

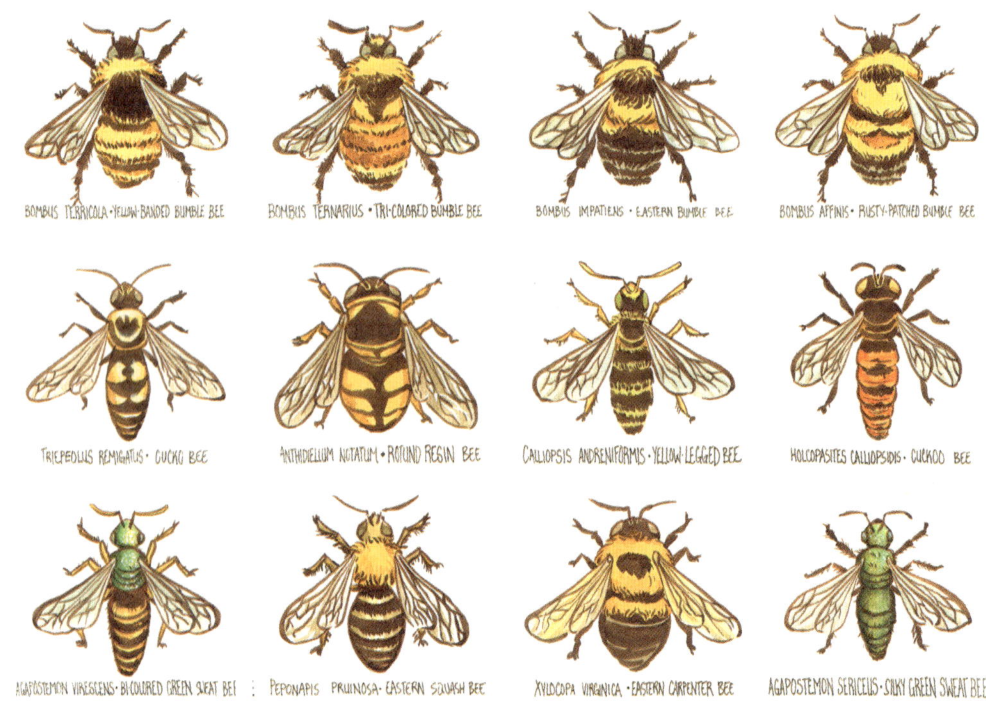

LIZ ANNA KOZIK

Liz Anna Kozik, *Native Bees of North America*, 2018. Pen and ink.

Can Agriculture Save Pollinators?

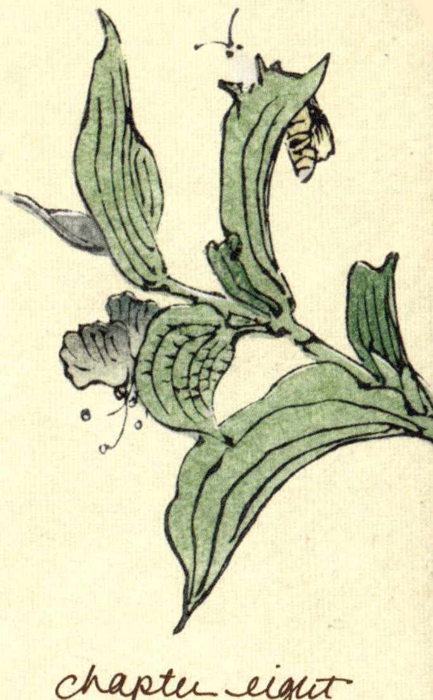

> How do we have the strength to not look away? And what do we see on the horizon if we do not make hard decisions and loving decisions?
>
> —TERRY TEMPEST WILLIAMS

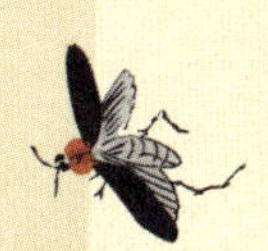

Heading southeast on a two-lane highway out of Madison, Wisconsin, the houses grow sparse and the sky expands, and you are soon surrounded by the most dominant crops grown in the Midwest: seed corn and soybeans. Recently, I drove this route through July-high fields on my way to Pied Beauty Farm in southern Wisconsin. I had been invited to the farm by its owners, Josh and Kerstin Mabie, to talk about the connection between bees and land.

The Mabies are not farmers by training. They are part of a trend of young professionals who have chosen to start farming on the side, or as a shift in vocation. They see this as a way to live closer to their ideals and help create a more sustainable way of growing food. After acquiring this land, they tilled up acres of soybeans and planted apple trees, hops, and prairie plants. They had plans to get honeybees in the spring. My first nonfiction book prompted them to invite me to take part in their educational lyceum series; they hoped some people would show up to eat some farm-fresh food together and talk about how to help pollinators.

I spotted the place from a distance, up on a small rise—a classic red barn, a small white farmhouse, and a huge garden filled with orange, red, and pink zinnias bordering a wild field of Dutch white clover, field grasses, and volunteer wildflowers—a distinct shift from even patterns of green. I was met on the gravel driveway by two enthusiastic ambassadors, the farmers' young children, who suggested I park in the grass. Their parents were in the kitchen preparing salads and pizzas for the guests. Slowly, people arrived: a teacher from the local school; a couple who had just purchased some goats; some college professors who knew Josh; a woman curious about attracting bees to her yard.

Pied Beauty Farm was the perfect setting to discuss issues facing pollinators. Looking out in one direction, we could see acres of tidy, even rows of soybeans, the verdant triumph of industrial agriculture, and in another, we saw apple trees, flowers, and a field of organic clover.

Our relationship to pollinators is complicated; humans need bees and other insects to fertilize much of our food and keep our ecosystems thriving, yet the way we grow this food is contributing to their decline. Our industrial agricultural landscapes are an ongoing threat to the health of pollinators, but changes in land use—like those on display at Pied Beauty Farm—may offer a saving grace.

If you spend much time with honeybees, you will be confronted with the complex interconnection of insects, flowers, and land. As a beekeeper, I knew this firsthand. Honeybees must live in a collaborative community to survive, and because they are intimately connected to the landscapes where they live, they need healthy forage and clean water. The colony lives in a small space, like a hive box or a tree, but because bees travel for pollen, nectar, and tree sap, the honeybee "home" is far larger than the hive itself. The honeybee world is a web, connecting them with the plants, water, trees, and soil of the surrounding landscape.

Unfortunately, human activity has made these connections much harder to maintain. Honeybees—and insects of all kinds—are dying at an unprecedented pace. I thought again of the study out of Germany that reported a 75 percent loss of flying insects over a period of twenty-seven years. Another, from the University of Vermont, shows specific evidence of wild bee decline in the United States. That we are in the midst of the Sixth Extinction. There are all kinds of reasons for this, including loss of habitat, climate change, disease, and pesticide use. One huge influence on the decline in pollinators, specifically, is the way much of the world farms.

The industrial agriculture model—also known as "conventional" farming—is built on a combination of monocropping and chemical inputs. As discussed in my chapter on bumblebees, this is the dominant model, especially in the United States. It's good for producing high volumes of specific products for an ever-growing global population, but it is designed to sustain only single crops. When these single crops need pollination, millions of hives of managed honeybees are shipped in. Often, there is not enough biodiversity in the landscape to sustain them once they pollinate the crop, so they are often fed high fructose corn syrup, which has been shown to be bad for their health. In most fields, there are no other flowering plants for forage.

A recent report by the United States Department of Agriculture estimates that, in 2018, soybean crops covered 89.6 million acres of farmland and grain crops, like corn, exceeded 81 million acres. "Producers planted 94 percent of the soybean acreage using herbicide resistant seed varieties, unchanged from 2017," the report explained. "Ninety-two percent of all corn acres planted in the United States are biotech varieties." That's a lot of area without biodiversity—and a lot of chemically dependent crops. Often, genetically modified crops are developed to be resistant to herbicides. Other crops are designed to have systemic pesticides embedded in the plant itself, many of which persist in the landscape even after the plant has decomposed.

The damaging effects of pesticides and herbicides are well known. Neonicotinoids, a class of pesticides that affect the neurological system of bees and other insects, contribute to pollinator decline. Imidacloprid and chlorpyrifos have been shown to impair the migratory patterns of seed-eating songbirds. An herbicide called glyphosate is known to disrupt the intestinal system of the honeybee. According to a 2016 study, "Since 1974 in the U.S., over 1.6 billion kilograms of glyphosate active ingredient have been applied, or 19% of estimated global use." While this works for increasing yields, it isn't good for pollinators.

Honeybee nervous systems are intricate and sensitive. Honeybees can read the vibrations of another bee dancing a map to a flower, in the dark, by touch alone. Honeybees' olfactory powers enable them to distinguish the scent of specific blossoms. Humans have even trained them to reliably detect diseases, drugs, and bombs. Their dexterity is crucial for grooming, for packing pollen onto their legs, for building honeycomb out of wax.

In my own life with honeybees, I have observed bees twitching uncontrollably after being poisoned. But often the effects of chemicals are not immediately observable. Some toxins disrupt insects' gut bacteria, and damage from neonicotinoids happens incrementally over time. It is excruciating to imagine these gentle, sensitive insects suffering from a neurotoxin, knowing the implications of their sensory impairment. The result is clear: for pollinators, the monocropping system, with its intensive chemical inputs, is not working well.

It puzzled me why people who work with land for a living—who understand its importance more than many of the rest of us—would use these products and farm in this way. I wanted to talk with a conventional farmer who could help me understand the conventional farm story, so I contacted an old family friend, Mel Gratton.

Gratton is from the rolling hills of northern Illinois, a couple of hours due west from Chicago, and has been in the farming business since he was a boy. He and his brother farm six hundred acres that have been in the family for more than 130 years. They raise corn, soybeans, and beef cattle. Gratton is in the business because he loves farming—planting, raising calves, and experiencing, as his wife Vickie, a retired nurse, reminds him, the "ritual of renewal" each spring. He talked enthusiastically about how farming had evolved in his seventy years. The transformations directly related to technological advances and economic pressures.

Gratton emphasized that farming used to be very hard work, physically as well as mentally. His grandparents pulled a plow with horses. His father's first tractor did not have power steering. Much of the farm work was necessarily done by hand. Gratton remembers the physical exhaustion he and his brother experienced after heaving square hay bales into the barn all day. "Today," he marveled, "a tractor can be driven by GPS, and a combine can map changing yields as it moves across the field. . . . The farmer can sit inside an air-conditioned cab and doesn't have to deal with climate conditions, like dust or cold, or biting insects."

He sees both the positive and negative aspects of this. "You can farm all day now and go home and still have a life, because you aren't completely drained. It's more of a business now, rather than a lifestyle," he says, "but that business supports a different kind of lifestyle."

Success, with this method of farming, has a lot to do with scale. The equipment and technology are not cheap, so a farmer needs a lot of acreage to support them. A good harvest from those acres needs to be reliable; many farmers are in deep debt. "[Farmers] can't afford to switch to something like specialty market vegetables on a whim," Gratton said. "They are pretty committed to commodity crops like soy, corn, or wheat when they have invested in this kind of equipment." So how do you ensure the success of vast fields of one plant? The answer comes in the advances in crop science.

Gratton explained to me how, before the introduction of Bt corn, a genetically modified plant, the European corn borer would burrow through a corn stock and end up in the ear of corn, making a pathway for pathogens and decreasing crop yields. By filling the corn with a bacterium that would kill the insect, farmers had more successful corn plants. "Roundup Ready" crops were another invention. These crops are immune to the herbicides used to kill everything around the main crop, decreasing competition. Gratton believes that crop science and technology have made it easier to feed more people with less work and that this is a very good thing.

When he explained it this way, his logic and his methods made sense. But still, I had reservations. I shared a story with Mel and Vickie about how helpless I felt when my daughter and I were sprayed by a crop duster, how scared I was when she grew sick afterward and one of the lymph glands in her neck grew to the size of a mango. Gratton paused and nodded. He agreed that not all chemicals are good ones.

Gratton was clearly familiar with the argument against pesticides and herbicides. "I am not against organic farmers," he said. He thinks there is room for all kinds of farmers and styles of farming, but he says, "Everything is not so black and white."

At the end of our conversation, Gratton talked about his commitment to stewardship of the land. He believes the land is a resource that he is using for his lifetime, and he says he hopes to improve it to the best of his ability. He practices no-till farming, for example, which helps keep land from eroding. He leaves corn stalks in the field so that they can decompose and return organic matter to the soil. He has areas in his land that are covered with forest, where many creatures can thrive. "This ethic of conserving the land for the future puts a tremendous pressure on everything I do," he said.

I left our conversation with a new appreciation for the evolution of farming and a deeper respect for Gratton, but I was still thinking about the bees and other insects—the foundation of our ecosystems, the pollinators of our crops, creatures I truly love. I sensed that many farmers had not chosen to use these pesticides and herbicides knowing that they would be bad for honeybees—or humans. They chose them because they were a solution to problems in production, and, like DDT, many of these chemicals were seen as a boon at first. Now we have more information, but the system we've created is hard to step out of.

After visiting Gratton, I drove out to the Driftless area in Wisconsin. "Driftless," they call this country, because the glaciers did not smooth it out. Driving into the Driftless, the roads get curvy and narrow as the hills multiply. Creeks and rivers serpentine through the valleys. Geographically speaking, it seems to make sense that farmers interested in smaller farms and more diverse crops would gravitate here. Many hillsides were covered in wildflowers, brambles, and trees, a haven for native insects. And yet I still passed miles of soy and corn.

The Thimmesch Family Farm is at the top of a hill in the Kickapoo Valley watershed. The modest yellow house is home to Jason and Jennelle Thimmesch and their four children, Myra, Isaac, Genevieve, and Nanka. The Thimmesches have grit and fierce idealism. Before moving to the town of Viroqua, they lived in a teepee off the grid into a Wisconsin winter. When they first moved to Viroqua, they worked as farmhands for a small CSA—a community-supported agriculture farm. The children were still very young when they bought this place, "in a dilapidated state," from a struggling Amish family.

Part of the draw to Viroqua, for the Thimmesches, was a conversation and then an extended visit with Mark Shepard, who is known for his book *Restoration Agriculture: Real-World Permaculture for Farmers*. A workshop in biodynamic farming inspired them further. Biodynamic farming was developed by Rudolf Steiner, a scientist and philosopher who believed that the farm should be seen as a "single, self-sustaining organism that thrives through biodiversity, the integration of crops and livestock and the creation of a closed-loop system of fertility." He wrote extensively about his reverence for the honeybee and predicted a serious decline of bees around the world if we did not shift our mechanized view of agriculture and the natural world.

Jason said that he and Jennelle were filled with "romantic idealism," and their original vision was to go "all-in." When an opportunity arose to work on an organic farm in the area as farmhands, they took it. Within a few years they were building their own dream farm. They had milk cows, sheep, chickens, honeybees, and a wide range of organic vegetables. They did everything by hand in the house and in the field with no electricity and they heated with wood, all while homeschooling their four small children. The oldest children worked with their parents in the fields.

But agricultural fundamentalism wasn't working for them. "I couldn't live up to the 'Ma Ingalls' image," Jennelle explained. She decided to start her own business in the nearby town. So they rethought their farm dream. Jason explained that he now focuses on three main certified organic crops for income: garlic, brussels sprouts, and cilantro. He raises some sheep, fifteen laying hens, and a few steers to feed family and friends and create manure. While he still uses some biodynamic preparations for the fields like a nettle tea or a concoction made of equisetum or horsetail, he goes against some biodynamic principles by purchasing some manufactured organic treatments.

The permaculture influence can still be seen in a wooded section of his land that is covered in hazelnuts, apples, and American chestnuts. This acre is snuggled up next to a stand of maple trees, which produce syrup in the spring. There is even a small, unkempt vineyard tucked between two of the main fields. Gazing at his land, I saw a rich tapestry of organic plants, habitat for a wide range of insects and birds.

The Amish have also made an impression on Jason. He still does nearly everything without mechanization. This year, he's planted nine thousand brussels sprout plants by hand and sold thirty thousand bunches of cilantro to places like Whole Foods and HyVee. He hires neighbors, including several Amish women and a few high school students from down the hill, to help. Often seven to nine people work together in the fields. Because Jason is out there daily, he knows the soil and the plants and the animals intimately and can respond when he sees an issue arise.

I asked him about a wilder area, where I spotted milkweed and Queen Anne's lace. "Yes," he said, "I leave this for the bees and the butterflies!" He no longer keeps honeybees, but between his neighbor's hives and all of the mason bees and bumblebees around, pollinators are plentiful on his property.

After lunch, Jason, Isaac, and I loaded the van with some boxes of brussels sprouts to make deliveries. One of our stops is Harmony Valley Farm. Harmony Valley consists of approximately three hundred acres, and it grows a wide range of produce for its fifteen hundred CSA members. Harmony Valley has addressed the decline of pollinators in several ways beyond simply farming without agrochemical pesticides. The CSA educates its members and gives out "Pollinator Packs," which are collections of native plants, such as purple coneflower and blue sage, that are beneficial to these crucial insects.

I thought of the blueberry growers I met while doing research for my first book. They planted native flowers next to their berries and discovered that providing healthy forage for native bees and honeybees increased both their yields and the size of their fruit. By attracting "beneficial insects" who prey on blueberry pests, they also decreased their need to use pesticides. And I had just returned from Amsterdam, which had made the decision to plant organic flowers in all of the city's green spaces—including beside train stations and along highways—and has seen, as a result, a 45 percent increase in native bee diversity and an increase in honeybee health.

Beekeepers have a role to play, too. Tom Seeley's research has shown that honeybees are healthier when they are not intensively and chemically managed. We can use his findings to change our own practices to be less invasive and more respectful. And many entomologists, like Claudio Gratton, Jeremy Hemberger (as we saw in chapter 7), and Rufus Isaacs are studying how native bees can pollinate as efficiently, if not more efficiently, than honeybees. With a new vision of farming and land use, perhaps we wouldn't need to use honeybees as industrial tools. Maybe we can turn the tide of insect decline.

Before I left the Driftless area, Jason Thimmesch and I stood on his farm, watching his shy sheep. Jason became more philosophical. He talked about the life of the farm as a way to be a good steward of the land, to support his family, and to connect to his community—including pollinators. He is not sure whether his children will continue farming here, but he seems confident that someone will continue caring for this land.

I hope Jason is right. I hope that whoever farms this land in the future, if it remains a farm, will keep its balance of cultivation and wildness. There is so much evidence that if we strike this balance just right, both humans and nonhumans can benefit.

On the road home, I passed hillsides covered with goldenrod under slowly reddening maples. I passed fields of cornstalks, gilded in the late light. I thought again about Pied Beauty Farm, Josh and Kerstin Mabie and their beautiful idealism, the healthy landscape they are creating. I thought about Mel Gratton and his belief in stewardship. Answers to the problems our planet is facing are not straightforward; there is no simple fix. We need to work together to create a more sustainable plan for humans and nonhumans alike. The bees were gathering the last bits of available pollen and nectar in preparation for winter. Would they survive another one?

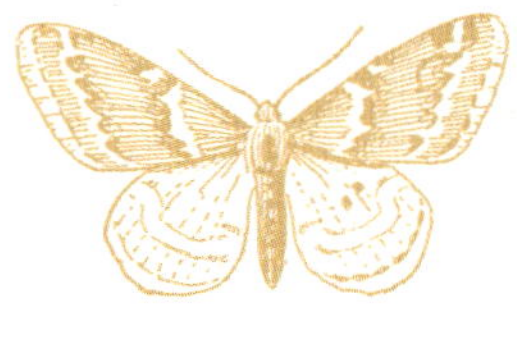

GALLERY EIGHT

Amy Spassov

The Inside Is the Outside

Not one, but also,
not two.
—NICK FLYNN

In August 2020, my father passed away after battling Parkinson's for many years. During the year following my father's death, I spent as much time as possible outside. I took long walks through the stands of oak and pine near my house, stood quietly in the middle of ever-changing prairies, floated in my canoe as the edges of days burned slowly down, colors draining until everything had the appearance of ash. Even during the winter, I trudged through fields of snow glowing blue in the moonlight, listening to the night, searching for answers, raw with loss. I felt more porous during this period of my life than I ever had before. I felt myself easily slipping into a space of transparency, born perhaps out of the confusion of grief as well as the desire to be held by the earth. I craved the solace of being among the many mortal things around me, many of them disappearing too fast, as my father had. What happened in those moments when my sense of self slipped is that my awareness of the other beings in these spaces became more prominent, and they seemed to become more comfortable with me. Curious owls and deer, sandhill cranes and common yellowthroats, bees and so many other insects, began to appear to me, almost like guides, as I moved slowly through the world. I wrote this poem during that period.

ON THE DAY AFTER YOU LEFT THIS WORLD

I floated out to the island
of bird bones, where
their long gone songs
now whisper in the cattails,
looking for solitude, solace,
but found instead
three cranes waiting
who let me join them
there on the shore,
their heads tipping
toward me, toward the
sounds of geese from
across the lake, toward
the jet plane flying overhead.
Night fell and we stayed—
all of us—cranes, crickets,

cattails, me with my broken body
breathing, and in the graying light
the breeze stroked
the cool waters of the lake,
the water lapping the mud
until all of it
was not separate, all of it
became one breath.

My revelation, which occurred more than once, was that we were not separate. The earth, even in death, tenderly cared for us, took us in, and changed us utterly. And in our lives, an openness to *not just thinking but feeling* the fact that we are a part of this miraculous, wildly diverse family of beings could bring one to a profound sense of peace and even a quiet kind of ecstasy.

The paintings of Amy Spassov instantly filled me with a sense of kindredness. Born in Portland, Oregon, and educated at the Seattle Art Institute, Spassov has shown her work in her own gallery in Bellingham, Washington, as well as in Los Angeles, Santa Fe, and in the United Kingdom. I could see in the incredibly vibrant, flower-filled figures that she must have had a revelation like mine. The dancing human bodies, frolicking deer bodies, and butterfly bodies were suffused with colorful images of petals, leaves, butterflies, delicate patterns, and even sometimes a tiny person. Nothing exists in these paintings without containing everything else.

When I eventually connected with Amy Spassov, she shared her own story of loss. During a year in her life when she was confronting infertility, she struggled with direction. In an article for *Creative Boom*, she wrote:

> Finding direction is complex. Hesitance and indecision rule the proceedings. That is unless, you have known the garden. Months were spent walking, observing and appreciating nature. It was here that I found direction, a means towards healing, acceptance and truth. The dust of these outings began to form on white panels. Nature, filtered and augmented, began to run onto the surfaces. Soon every piece would be entirely engulfed . . . I began extracting, ripping away the noise, the clutter, the doubt and all that was unnecessary to find clarity, self acceptance and a new direction.

In the garden, she found the healing she needed. Struggling with infertility, she found the fertility of the earth pulsing around her, and the garden entered her and poured out onto the canvas. The ideas of internality and externality, of separateness, are constantly challenged in her paintings. Where do we end and where does the outside begin? Her paintings and these questions brought to mind the poetry of Nick Flynn.

The poems in Nick Flynn's book *Blind Huber* tell the story of a beekeeper and his assistant and the bees they spend time studying. The character is based on a historical figure, who, despite his nearly complete blindness, observed and discovered many things about bee behavior. Some of the poems are about Huber, but many of the poems are written from the perspective of the bees themselves. As you read, you begin to see an indelible thread connecting all of the beings involved: honeybees, beekeepers, flowers.

Nick Flynn opens the book with a quote from the Vietnamese Buddhist monk Thich Nhat Hanh. The quote—"He already had the water, but he had to discover the jars"—references a story in which the Buddha, having attained enlightenment, was working to offer his insights in language to others. This quote offers the reader many things to think about. Certainly, it invokes a Buddhist way of knowing the world, but it also suggests that the beekeeper and the poet, and even perhaps the bee herself, have a kind of knowledge worth sharing. I would argue that the knowledge that arises from the poems, from the life of the hive, is that of interconnection, or interbeing, as Thich Nhat Hanh might say.

The concept arrives quickly in a poem called "Swarm," whose narrator is the swarm bees themselves, speaking as a whole. This choice of narrator is not a radical one if you recognize that a hive, because of each bee's dependence on the rest of the community, is considered to be a superorganism. The concept of this level of interdependence and cooperation applied to all beings is what is radical. The poem begins by addressing a human observer: "When you see us swarm . . . your mind / tries to make us one." "*I was born, /* you begin," the narrator says a few lines later, suggesting the "I" of the human ego, "& already each word / makes you smaller."

. . . Look at this field—

Cosmos. Lungwort. Utter each
& break
into a thousand versions of yourself.

. . . The answer is not one, but also
not two.

There is an endnote about these last two words written by Flynn. "Not two," he says, "is a line from what is considered the first Zen writing, according to M. P. Landis." "Not one, but also / not two" is exactly the beautiful contradiction that the bee world depends upon. The bees do not survive alone, and yet individuals fly to flowers, dance maps to others in the hive, care for the children, and remove the dead. Later, the voice of the honeybee speaks again of interdependence in the poem "Queen," when she says, "We pollinate the fields / because we are the fields." The poems do an extraordinary job of opening up a multispecies imaginary, to explore the Buddhist notion of interbeing, but they also invite us to see how these also apply to ourselves. Aren't we, too, individuals in these bodies, but like the bees, also inseparable from all that is around us? Are we not also breathing the air exhaled by plants? Are we not also sharing this generous earth?

When I asked Amy Spassov to tell me about what moved her to make her paintings, she shared a story about her personal sense of loss. The paintings featured in these pages were part of a healing process for her. The natural world began to fill the emptiness she had inside, and then the images made their way to the canvas. No amount of schooling, though she had plenty, could give her what this process did. It was more than a way to paint; it was a way to live. To be connected to the world around her.

After my father died, I opened the one beehive I had in my yard to find piles of dead bees. Somehow I was not surprised. A few months later, my cherished companion, my eight-year-old dog, died, too, suddenly. But spring arrived, and what had seemed dead grew green again and full of life. Dead trees, transformed by beetles and fungi into soil, sprouted new ferns. The seeds I placed in the cold ground grew into plants that blossomed. Wild bees arrived. Bumblebees and sweat bees. Migrating birds returned. Wintering birds continued their songs. No energy is ever destroyed, only transformed, scientists say. I find comfort in this. This cycle of which we are a part.

Not one, but also/ not two.

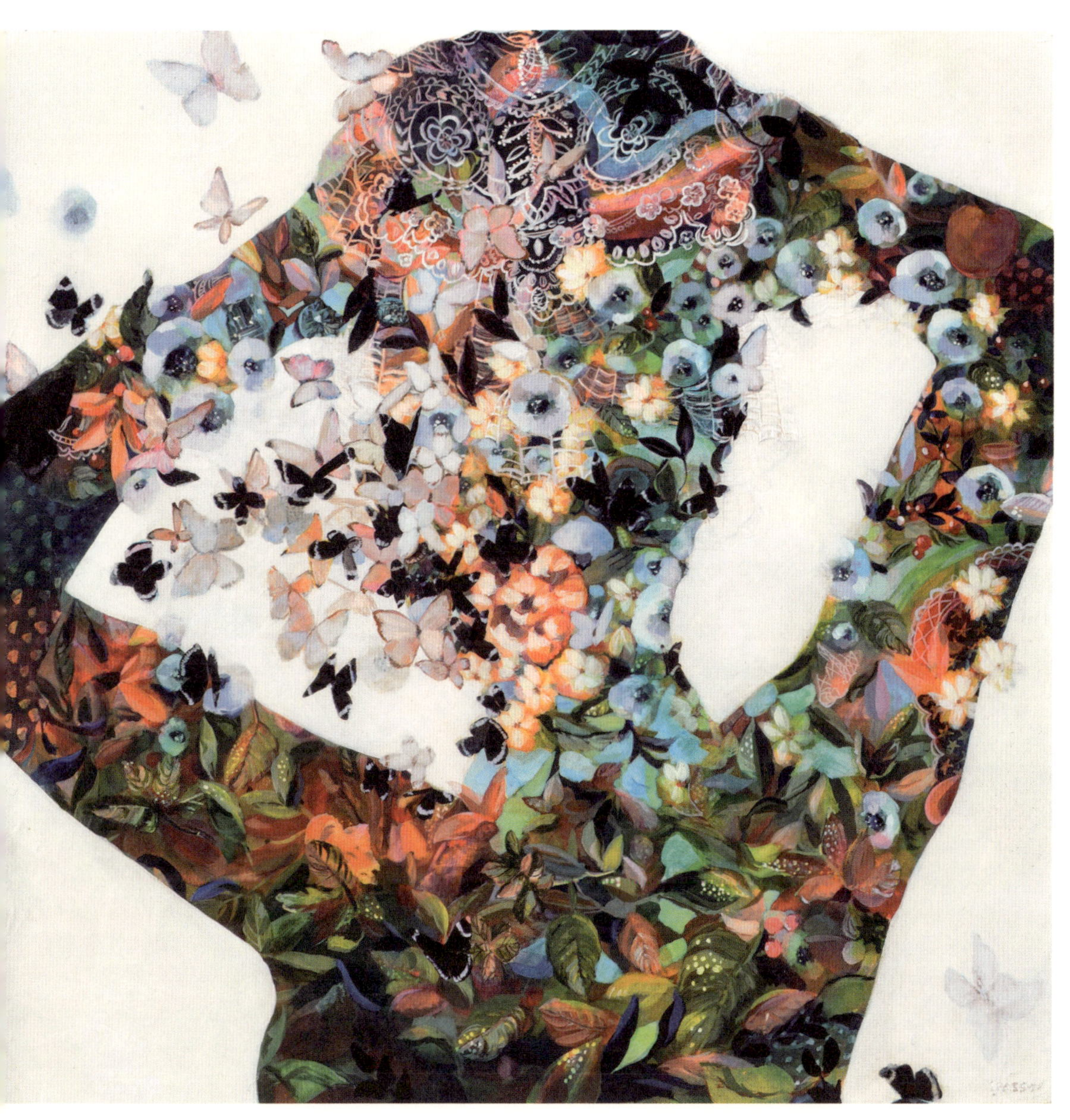

Amy Spassov, *Unveiling*, 2016. Mixed media on panel, 36 × 36 in. Photo: Hall Spassov.

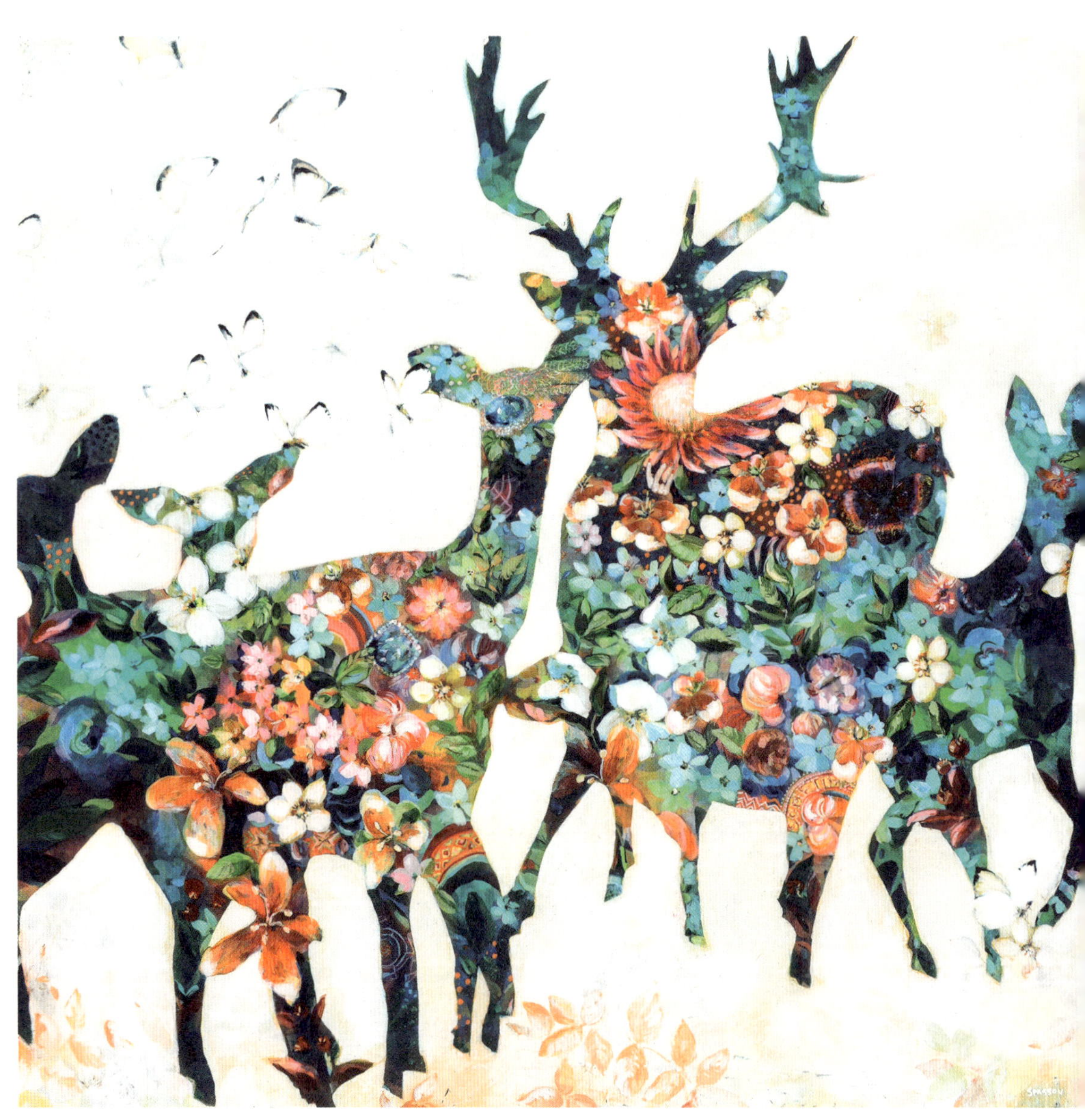

Amy Spassov, *There and Gone*, 2016. Mixed media on panel, 48 × 48 in. Photo: Hall Spassov.

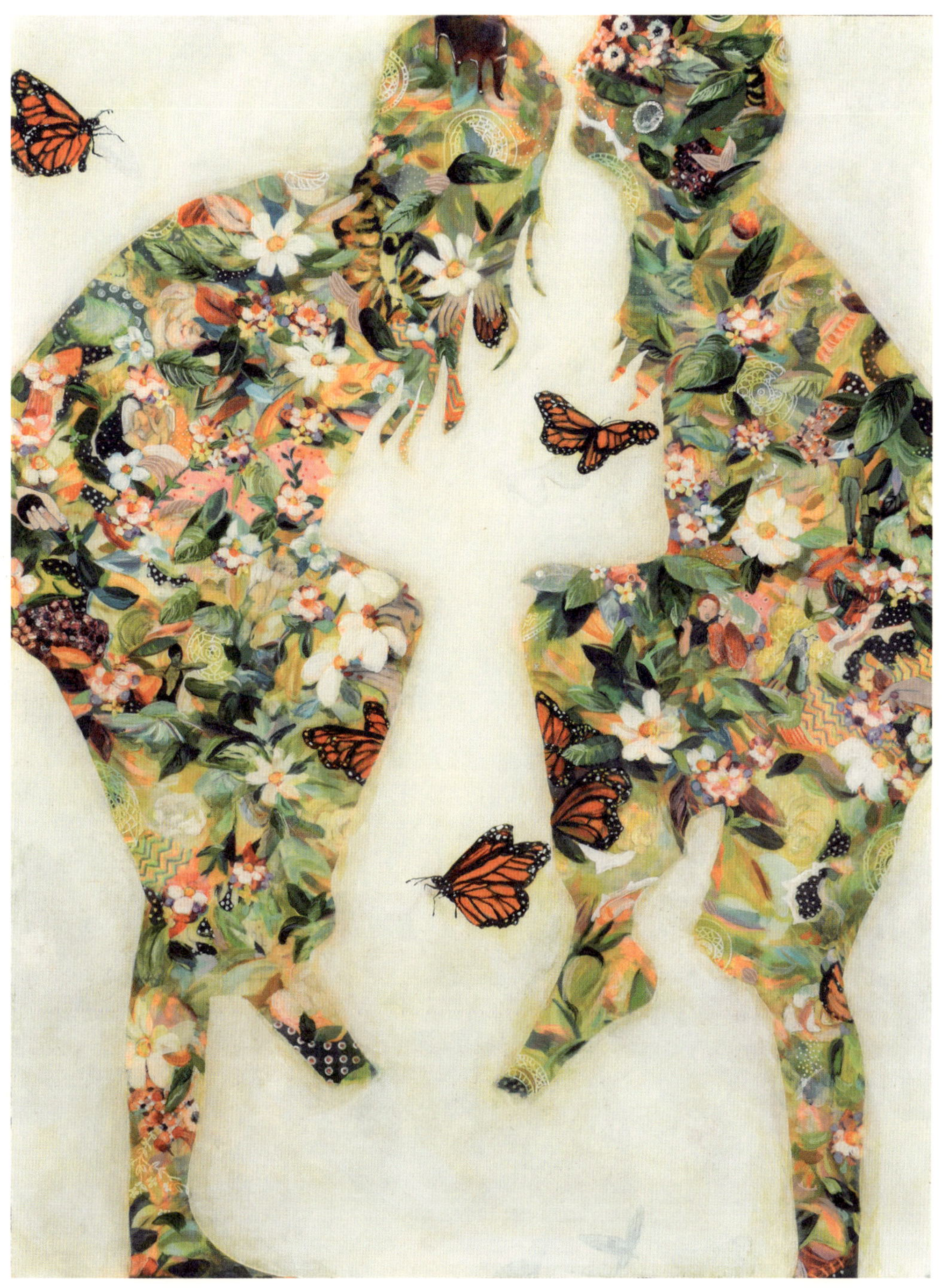

Amy Spassov, *The Undressing*, 2016. Mixed media on panel, 48 × 36 in. Photo: Hall Spassov.

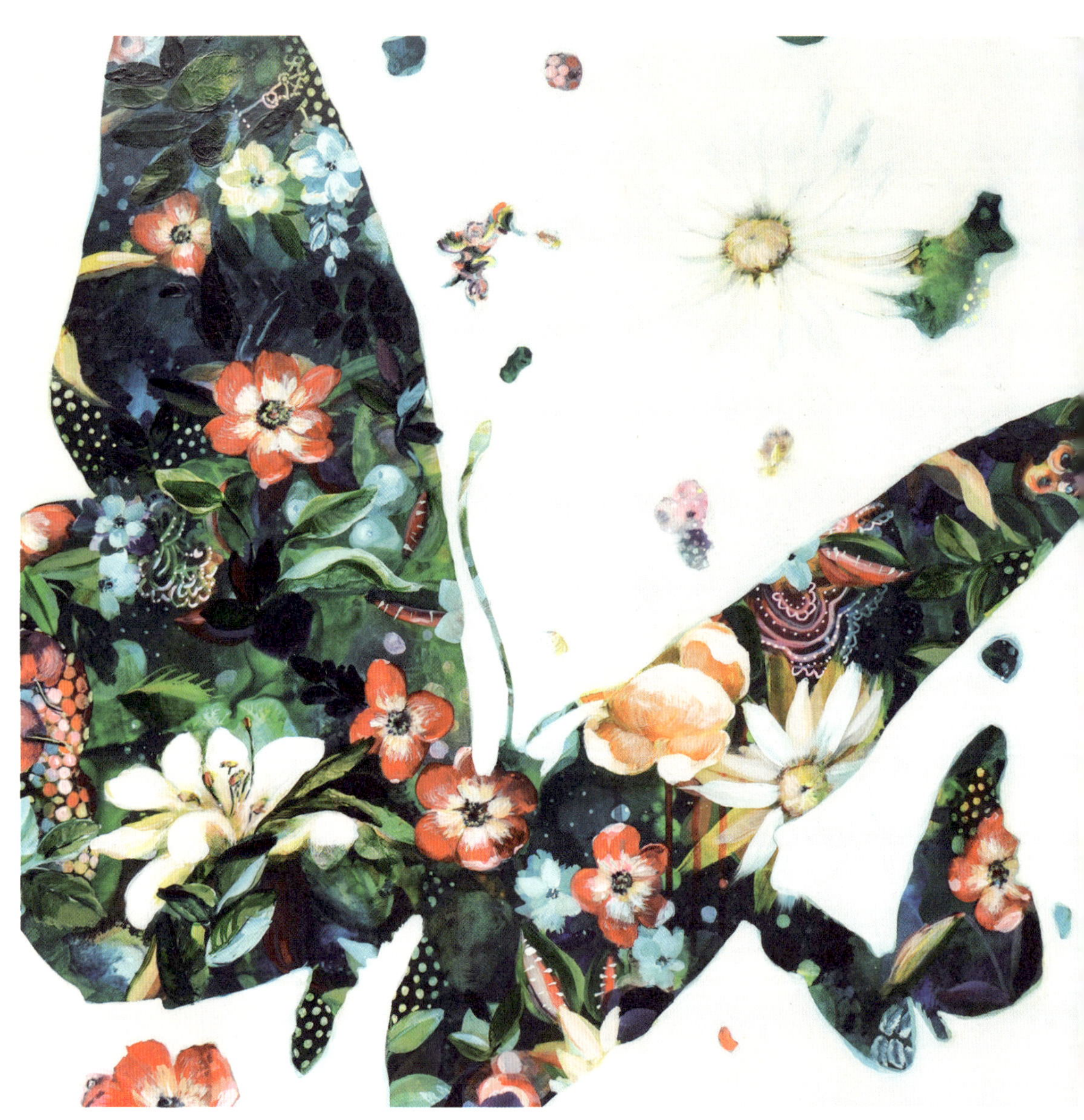

Amy Spassov, *New Growth*, 2016. Mixed media on panel, 24 × 24 in. Photo: Hall Spassov.

Afterword

COURAGE

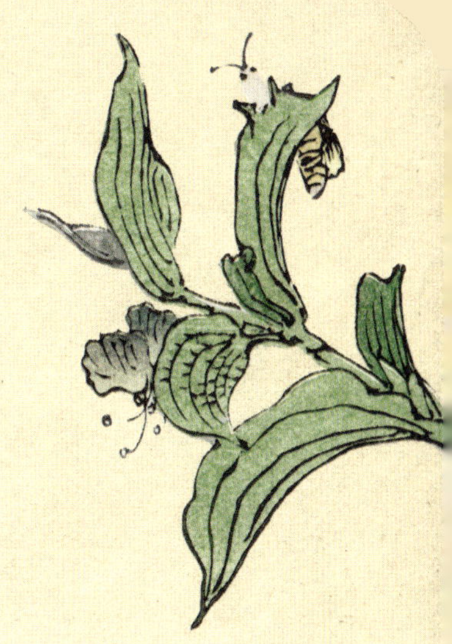

> Hope locates itself in the premises that we don't know what will happen and that in the spaciousness of uncertainty is room to act.
>
> —REBECCA SOLNIT

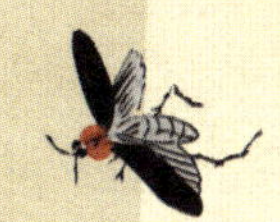
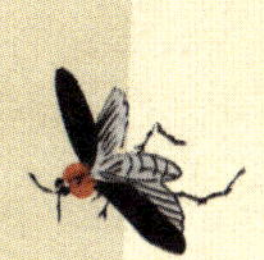

I teach college students in a university setting. For many years, teaching meant inviting people to think about things from a variety of perspectives, to think critically, to write clearly. During the last few years, however, I've added a few more things to my list. I started this book before the COVID-19 pandemic, but even then, I found students to be overwhelmed with a sense of sadness about the world they were inheriting. After living through the pandemic, many had learned ways of finding meaning, but the world ahead still looked daunting, especially for those students in environmental studies. One fall, I taught an eco-poetry class and decided that one of my goals was connection, connection with the poems, connection with each other, and connection with the earth. In the back of my mind was Aldo Leopold's belief: "When we see land as a community to which we belong, we may begin to use it with love and respect."

At first, going to school in person after the first wave of the pandemic was over was awkward; all of us had been staring at screens for over a year. We had to break through a kind of shyness to begin looking each other in the eye, to begin sharing what each of us thought. We read Robin Wall Kimmerer and her ideas about gratitude and reciprocity. We read Walt Whitman's poems about interconnectedness. We read J. Drew Lanham as he thought about how environmental justice and racial justice are inseparable. We read Juliana Sparr's moving *This Connection of Everyone with Lungs*. We read Pacyinz Lyfoung's "The Day I Learned to Speak My Grandmother's Tongue" in which the speaker has, like the poppies,

> Just awakened from their slumber
> By the baby chick with no feathers
> Hiding under the house board floor
> Waiting to teach the next generation
> That to live means to save the most vulnerable

"To live means to save the most vulnerable"! Who is the most vulnerable in any given situation? we began to ask ourselves, over and over. And slowly, as we read and discussed these poems and essays, we connected with each other on a heart level. We connected over the ideas and the possibilities we all yearned for. And then we connected with the planet. We were creating kinship. I had them sit by the lake under the trees and write about what they were sensing. I brought insects, some pinned and some alive, into the classroom so they could see the vast number of unique beetles and butterflies and to meet stick bugs and cockroaches from Madagascar in person. One of my students, a competitive weight lifter—arguably one of the strongest women, both physically and mentally, I have met—was terribly afraid of insects. That day she stepped through her fear and allowed a stick bug to crawl on her arm, and it was a revelation for her. I had them all write about a single maple leaf and read their responses out loud, and I will never think of a maple leaf in the same way. They wrote their own ecopoems. They were brightening as they shared their ideas and writings. They were being filled with feelings of loss and longing but also of wonder and curiosity. They were connecting with the planet as a whole. And where did this lead? It made them want to heal our broken hearts and our broken relationships with the land and with each other. It made them want to act.

But still, one day one of them surprised me with a response to a poem by Wendell Berry.

THE PEACE OF WILD THINGS

When despair for the world grows in me
and I wake in the night at the least sound
in fear of what my life and my children's lives may be,
I go and lie down where the wood drake
rests in his beauty on the water, and the great heron feeds.
I come into the peace of wild things
who do not tax their lives with forethought
of grief. I come into the presence of still water.
And I feel above me the day-blind stars
waiting with their light. For a time
I rest in the grace of the world, and am free.

The student said the poem offered her solace and a bit of hope, because most of the time she had none. She confessed that she did not want to bring children into this difficult world. I certainly respect anyone's choice to have children or not have children, but I sensed that maybe this had more to do with fear than with a desire to live child-free. Another person raised their hand and said they felt the same way. I put the question to the whole class. How many of them did not want to bring children into this world because it looked too dark out there? All of them. All of them raised their hands.

At this point I realized that it took more than connection and love. It also took courage. So I told them this story.

My mother loves to tell about the time she encountered a team of Belgians. For those who aren't familiar with Belgians, they are big, strong draft horses sometimes mistaken for Clydesdales, the horses made famous for pulling the Budweiser beer wagon. Because of their size and strength, some farmers still use Belgians for plowing and hauling. My mother makes pottery for a living, and for several years, she fired her mugs and bowls in a salt kiln she had built on a hill at the edge of a farm in northwest Illinois. Terry, the farmer who owned the land where she'd built her kiln, used Belgians to work his fields.

One late summer afternoon after a firing, my mother and her friend John were driving home through Terry's cornfields. Her cream-colored 1978 Chrysler station wagon, complete with fake wood paneling on the sides, barely fit between the walls of corn. The car crept along a single-lane dirt road only wide enough for a team of horses pulling a rake. Suddenly, on the top of the next rise, she saw the team of Belgians coming toward her on the same lane. My mother quickly realized that the horses were approaching at a full gallop, their heads tossed high in the air, their hooves thundering against the ground. Terry raced behind them on foot, waving his arms and shouting something they could not comprehend. John quickly climbed onto the hood of the station wagon, waving his hands in the air. "Whoa, big boys! Whoa!" he repeated, but they did not stop or even slow down. They kept coming. "They'll go around us," John shouted to my mother, who huddled in the driver's seat. But they kept coming, not slowing at all, and at the very last moment, John jumped out of their way, and my mother saw the horses raise their giant front hooves into the air and expose their white bellies as they tried to jump over the car.

As my mother curled into a ball on the passenger side, the mare's front feet crashed through the windshield, crushing the dashboard. The male horse had veered at the last minute to the right, and because they were connected by a harness, his momentum pulled his partner back out of the car. The glass tore her chest open before they both tumbled to the ground. Moments later, Terry arrived on the scene. He shouted to my mother and John to confirm that they were okay and then turned to the horses. They reared and whinnied wildly, blood everywhere. He grabbed onto the harness of his wounded horse near her neck and was flipped into the air, his body like a rag being shaken as she lifted off the ground, over and over, terrified. But Terry held on. He held on with one hand, and with the other he began tearing off his clothes and stuffing them into her wound. "**GET A VET**!" he shouted.

My mother ran the half mile to the farmhouse where there was a phone and called Doc Peterson to come out. By the time she made it back to the field, the horses were unhitched, both lying down. John and Terry hovered, wearing only their boxers, all of their clothes pressed into the wounds to stanch the blood. Doc Peterson eventually arrived and sewed up the horse's chest. That night, Terry and his wife slept with the horse in the field because she could not be moved. In the end, the horse lived, the people went back to their lives, and my mother's car was totaled.

The point of the story, as she tells it, is not primarily about the terrible violence, the clash of animals and technology, her folly for not knowing to move the car, or even about her survival. Rather, it is about Terry, the unflinching hero, who pushed past fear and hung on, his body being flung up and down, exhibiting the kind of courage love requires of us, the kind of endurance we must have.

I often wonder if I have that courage in me, and I like to think that I might be capable of doing what Terry did, but I have never had the occasion to save the life of a huge animal like that one. As a woman living in a medium-sized city in the United States of America in this particular decade in the twenty-first century, the number of opportunities I have had to embark on acts of physical heroism are few. But being a person living on this planet in this particular moment in history, when the earth is facing climate change, species extinction, environmental injustice, racism, and resource depletion, I do have the choice to enact a different kind of courage every day—which is the difficult choice of whether to love and to care about this planet and its inhabitants. Because loving others is risky.

Less than twenty-four hours after my daughter was born, we noticed her breaths becoming shorter and quicker. She was almost panting, her belly caving in each time the diaphragm worked to get more air. The doctors announced to me that she had a hole in each lung. Pneumothorax was the word for this particular kind of condition, a by-product of her botched C-section. The air building up in her chest cavity outside her lungs would soon collapse them. There was a terrible moment as they explained this, that I so hate to admit to, when I began to disconnect, to distance myself, to steel myself for the potential devastation of losing her.

But then I looked at her again, a little being with perfect hands and guileless eyes, with lungs so new to air, now gasping for survival: my child. I forced myself to turn toward that vulnerability that is love and stood fast in my attachment to her, despite the fear that burned through my body. Luckily, I had chosen, to the confusion of my nurses, not to take any kind of painkiller after my surgery, so I was wide awake. I watched as they inserted all kinds of tubes into her tiny body and flattened her in an X-ray machine nearly a hundred times. For several days, I stayed by her little glass cabinet in the neonatal intensive care unit. I did not sleep. I thought she needed someone to be holding on to, so I kept my hand on her and sang to her, over and over again, a little wordless lullaby my friend had once sung to me. A kind of mantra that kept me from coming unglued. My own incision seared when I moved—I was aware of it—but I knew I needed to stay. And eventually, thankfully, after many days, both of our wounds healed.

Loving something, caring for it, requires wading in with the whole heart splayed open. It is so easy to avoid caring, especially when it isn't for something like your child. And it's even harder when there is real work involved. My son, whose lifelong love of frogs has led me to meet frog experts like Ivan Lozano and frogs of many shapes and colors, and to learn the difference between the calls of bullfrogs and of spring peepers, told me recently that he thought it might be too depressing for him to continue studying them. Around the globe, many frogs are facing extinction due to a host of human-made threats, including habitat destruction, pesticide use, and climate change, as well as a deadly fungus called chytrid. He said he was afraid the frogs wouldn't make it. I reminded him of Lozano, of our friend Devin Edmonds in Madagascar who was successfully breeding endangered frogs, and of Brian Kubicki in Costa Rica who was rebuilding frog habitats. But inside I thought, of course, he's afraid. Aren't we all? No one wants to open themselves up to disappointment, to loss, to pain. So much simpler to remain detached, to avoid watching something suffer. I was certainly capable of it. But what if we all turn away?

I don't own a horse, but I do love bees. *Apis mellifera*, *Bombus impatiens*, *Xylocopa*, *Osmia avosetta*, *Apis dorsata*, Halictidae. . . . Like horses, the unique qualities of honeybees have been exploited by humans. Bees traverse our flowering fields, disseminate pollen, collect nectar. Some even magically produce honey. We depend upon their quiet labor. I love watching them floating in and out of their homes, delving into blossoms and covering their furry bodies with pollen, and stroking each other's faces with their antennae. I love the murmuring sound of a happy bee. But like frogs, as we now know, so many pollinators are struggling. We can't take them for granted. They are creatures who deserve to survive on this planet for their own sakes. They need help.

I have to admit getting overwhelmed, like my son, by all they were up against and thinking that any effort to change these trends would be futile. Sometimes I still feel this way. But for the past several years, I've been searching for the Terrys of the insect world. I wanted to learn about what was destroying insects and what was going to keep them alive. As I searched, I met a lot of heroes—entomologists, farmers, artists, conservation biologists, poets, coffee growers, botanists, firefighters—people I never dreamed of meeting. They are working hard, and often risking a lot, to keep our ecosystems vibrant and diverse. There are many different ways to care about something when it's not thriving, and sometimes healing takes much more than fixing a wound. Sometimes it takes changing our habits. Sometimes whole systems have to change. But the work of healing only happens when someone cares.

As I write these words, many beings on this planet are in danger and in need of care—insects, frogs, polar bears, salmon, elephants, and, of course, so many humans who are affected by droughts, who need clean water, who need clean air, who need safe neighborhoods. Their problems will not always be so obvious or immediate as a bleeding horse or our own suffering children, but when we recognize these problems, when we are faced with them, we will need the courage to refuse that temptation to shut down and walk away and instead to heroically open our hearts, to turn toward, and then to hold on.

In my part of the world, we have something very magical in the summer months. In the warm evenings, fireflies float up like tiny stars out of the fields as the crickets create a symphonic soundtrack. Not long ago, I slept in a cabin on the shore of the Wisconsin River. My friend and I spent the last hour of daylight paddling around sandbars and through smaller channels behind stands of trees as the water and the sky grew progressively more orange and then plum and then lavender. No motors, no sirens, no machines, just the sound of our paddles dipping in and out of the water—and the wild things. As the crepuscular hours crept past, the banks came alive with crickets, frogs, a beaver, a barred owl. Swallows swept low and fast over the water catching insects out of the air. Nightjars purred as they plunged high above the trees.

Once we had returned to the bank and retired to the porch to look into the ringing night, the stars began to pierce through the inky black, and soon a ballet of the glowing bodies of fireflies began, like tinier stars that had somehow dislodged themselves from the sky and now floated through the sultry air. Think of what a miracle that is. There are spaces like this left. I hope people generations from now will know the pure joy of seeing a firefly illuminating the tiny lantern of her body as she rises from the singing meadow into the miraculous evening air. For hours I stayed up, staring at them appearing and disappearing, the crickets and tree frogs still singing.

Notes

The notes below are keyed to the text by page numbers.

Introduction

1 *A 75 percent reduction of flying insects*: Caspar A. Hallmann, Martin Sorg, Eelke Jongejans, Henk Siepel, Nick Hofland, Heinz Schwan, Werner Stenmans, Andreas Müller, Hubert Sumser, Thomas Hörren, Dave Goulson, and Hans de Kroon, "More Than 75 Percent Decline Over 27 Years in Total Flying Insect Biomass in Protected Areas," *PLOS One* 12, no. 10 (2017): e0185809, https://doi.org/10.1371/journal.pone.0185809.

1 *The German study was not unique*: Jacob Mikanowski, "'A Different Dimension of Insect Loss': Inside the Great Insect Die-Off," *Guardian*, December 14, 2017, https://www.theguardian.com/environment/2017/dec/14/a-different-dimension-of-loss-great-insect-die-off-sixth-extinction.

5 *Her opus about DDT*: Rachel Carson, *Silent Spring* (Greenwich, CT: Crest Books, 1964).

7 *"Art is not in the service of science"*: Emily Arthur, *To Leave a Mark* (Madison, WI: University of Wisconsin Foundation, 2018).

9 *I open my classes by reading a poem*: Mary Oliver, "The Summer Day," in *New and Selected Poems* (Boston: Beacon Press, 1992), 1:94.

10 *Sometimes the cicadas are emerging early*: Chelsea Harvey, "Noisy Cicadas Are Emerging Earlier," *Scientific American*, March 22, 2021, https://www.scientificamerican.com/article/noisy-cicadas-are-emerging-earlier/.

Chapter 1

18 *Disappearance of the critically endangered palila*: Paul C. Banko, Peter T. Oboyski, John W. Slotterback, Steven J. Dougill, Daniel M. Goltz, Luanne Johnson, and Megan E. Laut, "Availability of Food Resources, Distribution of Invasive Species, and Conservation of a Hawaiian Bird Along a Gradient of Elevation," *Journal of Biogeography* 29 (2002): 789–808, https://digitalcommons.unl.edu/usgsstaffpub/639.

19 *Revisited forty-five specific sites*: Joan E. Ball-Damerow, Leithen K. M'Gonigle, and Vincent H. Resh, "Changes in Occurrence, Richness, and Biological Traits of Dragonflies and Damselflies (Odonata) in California and Nevada Over the Past Century," *Biodiversity and Conservation* 23 (2014): 2107–26, https://doi.org/10.1007/s10531-014-0707-5.

19 *One dragonfly can eat*: Sarah Zielinski, "14 Fun Facts About Dragonflies," *Smithsonian Magazine*, October 5, 2011, https://www.smithsonianmag.com/science-nature/14-fun-facts-about-dragonflies-96882693/.

23 *A scientist named Robert Hooke*: Robert Hooke, *Micrographia: or Some Physiological Descriptions of Minute Bodies Made by Magnifying Glasses with*

Observations and Inquiries Thereupon (London: Royal Society, 1665), http://www.academia.dk/MedHist/Biblioteket/Print/hooke_micrographia_plain.html.

24 *Hooke "was curator of experiments"*: Tom Lubbock, "The Flea (1665), Robert Hooke," *The Independent*, November 13, 2009, https://www.independent.co.uk/arts-entertainment/art/great-works/the-flea-1665-robert-hooke-1819454.html.

24 *Disgust holds our attention*: Sianne Ngai, *Ugly Feelings* (Cambridge, MA: Harvard University Press, 2005).

Chapter 2

32 *An ornithologist named Frank Chapman*: "History of the Christmas Bird Count," *Audubon*, January 8, 2018, https://www.audubon.org/conservation/history-christmas-bird-count.

33 *A riveting biodiversity report*: "UN Report: Nature's Dangerous Decline 'Unprecedented'; Species Extinction Rates 'Accelerating,'" United Nations Sustainable Development Goals blog, May 6, 2019, https://www.un.org/sustainabledevelopment/blog/2019/05/nature-decline-unprecedented-report.

33 *The study out of Germany*: Hallmann et al., "More Than 75 Percent Decline."

34 *"More biodiversity per square kilometer"*: Aristos Georgiou, "Is This Arachnid with the Face of a Bunny and Dog Morphed Together Cute or Terrifying?," *Newsweek*, November 6, 2018, https://www.newsweek.com/arachnid-face-bunny-and-dog-morphed-together-cute-or-terrifying-1202985.

34 *"Guarantees the right to live in peace"*: Sophie Moon, "Ecuador Grants Open-Pit Mining in One of the World's Most Biodiverse Areas," *Truthout*, April 1, 2018, https://truthout.org/articles/ecuador-grants-open-pit-mining-permits-in-one-of-the-world-s-most-biodiverse-areas/.

34 *Recent studies … using "cacao starch grains"*: Sonia Zarrillo, Nilesh Gaikwad, Claire Lanaud, Terry Powis, Christopher Viot, Isabelle Lesur, Olivier Fouet, et al., "The Use and Domestication of *Theobroma Cacao* During the Mid-Holocene in the Upper Amazon," *Nature Ecology and Evolution* 2 (2018): 1879–88, https://doi.org/10.1038/s41559-018-0697-x.

35 *Good for these finicky flowers*: Kofi Frimpong-Anin, Michael K. Adjaloo, Peter K. Kwapong, and William Oduro, "Structure and Stability of Cocoa Flowers and Their Response to Pollination," *Journal of Botany*, 2014, article ID 513623, https://doi.org/10.1155/2014/513623.

38 *"Two of Ecuador's five mines"*: Alexandra Valencia, "Ecuador Says Two Mines on Track to Producing Gold, Copper This Year," *Reuters*, February 28, 2019, https://www.reuters.com/article/us-ecuador-mining/ecuador-says-two-mines-on-track-to-producing-gold-copper-this-year-idUSKCN1QH2RT.

39 *Recent critique of tourism*: Van Badham, "After Everest, We Have to Rethink the Places We Are Loving to Death," *Guardian*, June 18, 2019, https://www.theguardian.com/commentisfree/2019/jun/19/after-everest-we-have-to-rethink-the-places-we-are-loving-to-death.

39 *Identifying the ones she finds*: Samuel Green, "Butterflies" (2017), Poetry

Foundation, https://www.poetryfoundation.org/poems/147795/butterflies-5b884902ce425.

40 *Doubted the power of the "rights of nature"*: Mary E. Whittenmore, "The Problem of Enforcing Nature's Rights Under Ecuador's Constitution: Why the 2008 Environmental Amendments Have No Bite," *Pacific Rim Law and Policy Journal* 20, no. 3 (2011): 659–91.

40 *"Policy frameworks that place humans in context"*: "Landmark Ruling Blocks Mining in Ecuadorian Forest, Citing Rights of Nature," E360 Digest, *YaleEnvironment360*, December 3, 2021, https://e360.yale.edu/digest/landmark-ruling-blocks-mining-in-ecuadorian-forest-citing-rights-of-nature.

42 *Ceiba was "promoting forest protection"*: "Coastal Conservation Corridor," Ceiba Foundation, https://ceiba.org/conservation/dry-forests/coastal-conservation-corridor/.

42 *At least 132 species of stingless bees*: Patricia Vit, Silvia R. M. Pedro, Favian Maza, Virginia Meléndez Ramírez, and Viviana Frisone, "Diversity of Stingless Bees in Ecuador, Pot-Pollen Standards, and Meliponiculture Fostering a Living Museum Meliponini of the World," in *Pot-Pollen in Stingless Bee Melittology*, edited by Patricia Vit, Silvia R. M. Pedro, and David W. Roubik (Cham, Switzerland: Springer, 2018), 207–28, https://doi.org/10.1007/978-3-319-61839-5_15.

44 *Keep the bees and the coffee thriving*: Pablo Imbach, Emily Fung, Lee Hannah, Carlos E. Navarro-Racines, David W. Roubik, Taylor H. Ricketts, Celia A. Harvey, et al., "Coupling of Pollination Services and Coffee Suitability Under Climate Change," *Proceedings of the National Academy of Sciences* 114, no. 39 (September 2017): 10438–42, https://doi.org/10.1073/pnas.1617940114.

45 *Imagining what the transformation might be like*: James Crews, "Monarch," in *Bluebird: Poems* (Brattleboro, VT: Green Writers Press, 2020).

46 *The caterpillar digests itself*: Ferris Jabr, "How Does a Caterpillar Turn into Butterfly?," *Scientific American*, August 10, 2012, https://www.scientificamerican.com/article/caterpillar-butterfly-metamorphosis-explainer/.

46 *"She wanted to slow down"*: Callum Angus, "The Moonsnail," *West Branch* 90 (Spring–Summer 2019).

48 *Ovid begins his epic retelling*: Ovid, *Metamorphosis*, trans. David R. Slavitt (Baltimore: Johns Hopkins University Press, 1994).

Chapter 3

54 *Flower exports bring approximately $1.5 billion*: Benjamin Rau and Lady Gomez, "Sales of Colombian Flowers Fall During Covid-19 Crisis," USDA Foreign Agricultural Service / GAIN Attaché Report, May 14, 2020, https://www.fas.usda.gov/data/colombia-sales-colombian-flowers-fall-during-covid-19-crisis. For more on Colombian orchids, see Oscar Alejandro Perez Escobar and Meryl Westlake, "Colombia's Most Interesting Orchids," Kew Gardens, https://www.kew.org/read-and-watch/colombias-interesting-orchids, and Marta Kolonowska, "The Orchid Flora of the Colombian Department of Valle del Cauca," *Revista Mexicana de Biodiversidad* 85, no. 2 (2014) 445–62, https://www.sciencedirect.com/science/article/pii/S1870345314707729. On Andean ecosystems, see Nicolas

Hazzia, Juan Sebastián Moreno, Carolina Ortiz-Movliav, and Rubén Darío Palacio, "Biogeographic Regions and Events of Isolation and Diversification of the Endemic Biota of the Tropical Andes," *Proceedings of the National Academy of Sciences* 115, no. 31 (2018): 7985–90, https://doi.org/10.1073/pnas.1803908115.

55 *"The Greeks named this plant"*: Lucy Hooper, ed., *The Lady's Book of Flowers and Poetry: To Which Are Added a Botanical Introduction, a Complete Floral Dictionary; and a Chapter on Plants in Rooms* (New York: J. C. Riker, 1842), 130.

55 *"Made into a remedy for sick elephants"*: "The Orchid in the Human Imagination," Center for Research in Reproduction and Contraception, University of Washington, https://depts.washington.edu/popctr/orchids.htm.

55 *"Ancient Chinese called eternal friendship 'lan jiao'"*: "Poetic Beauty: 10 Most Significant Flowers in China," *China Daily*, March 1, 2017, http://www.chinadaily.com.cn/interface/zaker/1143605/2017-03-01/cd_28388933.html.

55 *The Japanese poet Izumi Shikibu*: Jane Hirshfield, trans., with Mariko Aratani, *The Ink Dark Moon* (New York: Vintage Books, 1986).

56 *A single Shenzhen Nongke orchid*: Melissa Breyer, "8 of the Most Expensive Flowers in the World," *Treehugger*, February 2022, https://www.treehugger.com/most-expensive-flowers-in-the-world-4864230.

56 *The illegal removal of native orchids*: Maximo Anderson, "The Lure of Wild Orchids Persists in Colombia," Mongabay.com, November 2, 2017, https://news.mongabay.com/2017/11/the-lure-of-wild-orchids-persists-in-colombia/.

57 *A study published in "PLOS One"*: Hallmann et al., "More Than 75 Percent Decline."

57 *Another study, published in "Biological Conservation"*: Francisco Sánchez-Bayo and Kris A. G. Wyckhuys, "Worldwide Decline of the Entomofauna: A Review of Its Drivers," *Biological Conservation* 232 (April 2019): 8–27, https://www.sciencedirect.com/science/article/pii/S0006320718313636.

57 *A recent PNAS report*: David L. Wagner, Eliza M. Grames, Matthew L. Forister, May R. Berenbaum, and David Stopak, "Insect Decline in the Anthropocene: Death by a Thousand Cuts," *Proceedings of the National Academy of Sciences* 118, no. 2 (January 2021): e2023989118, https://doi.org/10.1073/pnas.2023989118.

58 *Over 220,000 people were killed*: Associated Press, "Colombian Conflict Has Killed 220,000 in 55 Years, Commission Finds," *Guardian*, July 25, 2013, https://www.theguardian.com/world/2013/jul/25/colombia-conflict-death-toll-commission.

60 *Danger of being an environmental defender*: Rob Nixon, "Defending Tomorrow Today," *Edge Effects*, August 6, 2020 (updated October 9, 2021), https://edgeeffects.net/defend-environmental-defenders.

60 *"Nature's balance was created by diversity"*: Andrea Wulf, *The Invention of Nature: The Adventures of Alexander Von Humboldt, the Lost Hero of Science* (London: John Murray, 2015), 125.

61 *Dozens of strange orchid/pollinator relationships*: Joe Meisel, *Orchids of Tropical America: An Introduction and Guide* (Ithaca: Comstock, 2014).

63 *A poem by R. Snow*: Hooper, *Lady's Book of Flowers*, 132.

62 *The mycorrhizal networks that connect plants*: Richard Grant, "Do Trees Talk to Each Other?," *Smithsonian Magazine*, March 2018, https://www.smithsonianmag.com/science-nature/the-whispering-trees-180968084/.

Chapter 4

72 *A multi-billion-dollar enterprise*: US Fish and Wildlife Service, "Combating Wildlife Trafficking," https://www.fws.gov/program/combating-wildlife-trafficking. See also Katherine Lawson and Alex Vines, "Global Impacts of the Illegal Wildlife Trade," Chatham House, February 2014, https://www.chathamhouse.org/sites/default/files/public/Research/Africa/0214Wildlife.pdf.

83 *"Chemical substances used to prevent"*: Michael C. R. Alavanja, "Pesticides Use and Exposure Extensive Worldwide," *Reviews on Environmental Health* 24, no. 4 (2009): 303, https://www.ncbi.nlm.nih.gov/pmc/articles/PMC2946087/.

Chapter 5

88 *"Dirt—or more precisely, soil"*: Paul Bogard, *The Ground Beneath Us: From the Oldest Cities to the Last Wilderness, What Dirt Tells Us About Who We Are* (New York: Little, Brown, 2017), 96–97.

90 *One female can produce two hundred eggs*: Oklahoma State University Extension, "Boll Weevil," https://extension.okstate.edu/programs/digital-diagnostics/insects-and-arthropods/boll-weevil-anthonomus-grandis/.

90 *A report by the entomologist Dominic Reisig*: Matt Shipman, "The Boll Weevil War, or How Farmers and Scientists Saved Cotton in the South," North Carolina State University News, May 17, 2017, https://news.ncsu.edu/2017/05/boll-weevil-war-2017/.

91 *The weevil built resistance to pesticide after pesticide*: Ron Smith, "Boll Weevil in Alabama," Encyclopedia of Alabama, January 22, 2008, last updated March 30, 2023, http://encyclopediaofalabama.org/article/h-1436.

91 *One-third of all the pesticides used*: Shipman, "Boll Weevil War."

91 *Cotton was a "child-labor-intensive crop"*: Richard B. Baker, John Blanchette, and Katherine Eriksson, "Long-Run Impacts of Agricultural Shocks on Educational Attainment: Evidence from the Boll Weevil," *Journal of Economic History* 80, no. 1 (2020): 136–74, https://doi.org/10.1017/S0022050719000779.

91 *A less violent South*: James J. Feigenbaum, Soumyajit Mazumder, and Cory B. Smith, "When Coercive Economies Fail: The Political Economy of the US South After the Boll Weevil," National Bureau of Economic Research, May 2020, https://www.nber.org/papers/w27161.

91 *Credit the boll weevil . . . for the Great Migration*: Shipman, "Boll Weevil War."

92 *"Patton may have been performing this tune"*: Robert K. D. Peterson, "Charley Patton and His Mississippi Boweavil Blues," *American Entomologist* 53, no. 3 (Fall 2007): 143, https://doi.org/10.1093/ae/53.3.142. For more on the history of the boll weevil, see James

C. Giesen, "The Truth About the Boll Weevil," *Mississippi History Now*, March 2015, http://www.mshistorynow.mdah.ms.gov/articles/391/the-truth-about-the-boll-weevil, and Randal L. Hall, review of *Boll Weevil Blues: Cotton, Myth, and Power in the American South*, by James C. Giesen, EH.net, January 2013, https://eh.net/book_reviews/boll-weevil-blues-cotton-myth-and-power-in-the-american-south/.

93 *Farmers in the community began to "heed the advice"*: Ben Berntson, "The Boll Weevil Monument," Encyclopedia of Alabama, August 10, 2009, last updated March 27, 2023, http://encyclopediaofalabama.org/article/h-2384.

94 *Carver's ingenuity laid the foundation*: Monica M. White, *Freedom Farmers: Agricultural Resistance and the Black Freedom Movement* (Chapel Hill: University of North Carolina Press, 2019), quotations from 46.

94 *Snare boll weevils using their own pheromones*: Smith, "Boll Weevil in Alabama."

96 *"About ten times the weight of a warbler"*: Bernd Heinrich, *Life Everlasting: The Animal Way of Death* (Boston: Mariner Books, 2013).

97 *"The smallest sprout shows"*: Walt Whitman, "Song of Myself" (section 6), from *Leaves of Grass* ([Philadelphia]: David McKay, 1891), https://poets.org/poem/song-myself-6-child-said-what-grass.

Chapter 6

105 *"The tallgrass prairie of central North America"*: Peter H. Ravena and David L. Wagner, "Agricultural Intensification and Climate Change Are Rapidly Decreasing Insect Biodiversity," *Proceedings of the National Academy of Sciences* 118, no. 2 (2021): e2002548117, https://doi.org/10.1073/pnas.2002548117.

106 *Labeled a Superfund site*: Environmental Protection Agency, "Superfund Site: Joliet Army Ammunition Plant (Load-Assembly-Packing Area), Joliet, IL," https://cumulis.epa.gov/supercpad/cursites/csitinfo.cfm?id=0501170.

107 *The prairie plants*: Doug Ladd and Frank Oberle, *Tallgrass Prairie Wildflowers: A Field Guide* (Helena, MT: Falcon, 1995). For more on native plants and pollinators, see Gordon W. Frankie, Robbin W. Thorp, Rollin E. Coville, and Barbara Ertter, *California Bees and Blooms: A Guide for Gardeners and Naturalists* (Berkeley, CA: Heyday, 2004), and Xerces Society, *100 Plants to Feed the Bees: Provide a Healthy Habitat to Help Pollinators Thrive* (North Adams, MA: Storey, 2016).

108 *Over a billion tons of TNT*: US Forest Service / USDA, "The Joliet Army Ammunition Plant," Midewin National Tallgrass Prairie: History and Culture, https://www.fs.usda.gov/detail/midewin/learning/history-culture.

108 *Andrew Mahlstedt's arresting essay*: Andrew Mahlstedt, "On Edge in the 'Devil's Gardens' of Bosnia," *Edge Effects*, January 27, 2015, last updated October 12, 2019, https://edgeeffects.net/devils-gardens-of-bosnia/.

108 *The creation of this thriving tallgrass prairie*: "Midewin National Tallgrass Prairie," US Forest Service / USDA, https://www.fs.usda.gov/midewin.

111 *The word Midewin*: "Meaning of Midewin," Illinois State Museum,

https://www.museum.state.il.us/exhibits/midewin/meaning.html.

114 *The protection of the California gnatcatcher*: Jim Tranquada, "Emily Arthur—Endangered," Moore Laboratory of Zoology, Occidental College, February 12, 2016, https://moorelab.oxy.edu/news/emily-arthur-endangered-oxy-march-2-april-9.

114 *"Endemic to the Palos Verdes Peninsula"*: Richard A. Arnold, "Decline of the Endangered Palos Verdes Blue Butterfly in California," *Biological Conservation* 40, no. 3 (1987): 203–17, https://doi.org/10.1016/0006-3207(87)90086-3.

117 *"And yet a particular way of existence was gone"*: Lydia Millet, *How the Dead Dream* (New York: First Mariner Books, 2009), 166.

Chapter 7

122 *"We abuse land"*: Aldo Leopold, *A Sand County Almanac* (New York: Oxford University Press, 2020), xxii.

123 *Bumblebees are . . . in serious decline*: Peter Soroye, Tim Newbold, and Jeremy Kerr, "Climate Change Contributes to Widespread Declines Among Bumble Bees Across Continents," *Science*, February 7, 2020, https://science.sciencemag.org/content/367/6478/685.

125 *We spent $105 billion on our lawns*: Tim Root, "Ditching Lawns Could Help Your Backyard Thrive," *Washington Post*, June 30, 2021, https://www.washingtonpost.com/climate-solutions/2021/06/30/climate-friendly-backyard/.

125 *Homeowners spray their lawns*: Margaret Renkl, "America's Killer Lawns," *New York Times*, May 18, 2020, https://www.nytimes.com/2020/05/18/opinion/lawn-pesticides-insect-extinction.html.

128 *"According to the Ojibwe"*: Huron Smith, "Ethnobotany of the Ojibwe Indians," *Bulletin of the Public Museum of the City of Milwaukee* 4, no. 3 (1932): 327–525, plates 46–77, http://blogs.nwic.edu/briansblog/files/2013/02/Ethnobotany-of-the-Ojibwe-Indians.pdf.

128 *The cranberry is known . . . to be high in antioxidants*: Sarah Whitman-Salkin, "Cranberries, a Thanksgiving Staple, Were a Native American Superfood," *National Geographic*, November 28, 2013, https://www.nationalgeographic.com/science/article/131127-cranberries-thanksgiving-native-americans-indians-food-history.

129 *"Honey bees are the most effective pollinators"*: Marla Spivak, "What Can You Do to Improve Cranberry Pollination?," UW Fruit Program, https://fruit.wisc.edu/wp-content/uploads/sites/36/2011/05/What-Can-You-Do-to-Improve-Cranberry-Pollination.pdf.

129 *Jeremy Hemberger's bumblebee research*: Jeremy Hemberger, "Saved by the Pulse? Separating the Effects of Total and Temporal Food Abundance on the Growth and Reproduction of Bumble Bee Microcolonies," *Basic and Applied Ecology*, April 2020.

131 *"Recognition of the personhood of other beings"*: Robin Wall Kimmerer, "Returning the Gift," *Minding Nature* 7, no. 2 (2014), https://grateful.org/resource/returning-the-gift/.

Chapter 8

139 *A recent report by the United States Department of Agriculture*: Jim Barrett, "USDA Reports Soybean, Corn Acreage Down," USDA National Agricultural Statistics Service, June 29, 2018,

https://www.nass.usda.gov/Newsroom/archive/2018/06-29-2018.php.

139 *According to a 2016 study*: C. M. Benbrook, "Trends in Glyphosate Herbicide Use in the United States and Globally," *Environmental Sciences Europe* 28 (2016), https://doi.org/10.1186/s12302-016-0070-0.

142 *A "single, self-sustaining organism"*: Biodynamic Demeter Alliance, "Who Was Rudolf Steiner?," https://www.biodynamics.com/steiner.html.

146 *"Finding direction is complex"*: Katy Cowan, "I've Known the Garden: New Mixed Media Art Inspired by Nature by Amy Spassov," *Creative Boom*, May 27, 2016, https://www.creativeboom.com/inspiration/ive-known-the-garden-new-mixed-media-art-inspired-by-nature-by-amy-spassov/.

147 *Nick Flynn opens the book*: Nick Flynn, *Blind Huber* (Minneapolis: Graywolf Press, 2002).

Afterword

153 *"When we see land as a community"*: Leopold, *Sand County Almanac*, xxii.

154 *"Just awakened from their slumber"*: Pacyinz Lyfoung, "The Day I Learned to Speak My Grandmother's Tongue," *Split This Rock*, May 21, 2019, https://www.splitthisrock.org/poetry-database/poem/the-day-i-learned-to-speak.

155 *"When despair for the world grows in me"*: Wendell Berry, "The Peace of Wild Things," in *Openings: Poems by Wendell Berry* (New York: Harcourt, Brace and World, 1968).

Resources and Inspiration

Carson, Rachel. *Silent Spring*. Greenwich, CT: Crest Books, 1964.

Dillard, Annie. *Pilgrim at Tinker Creek*. 1974. New York: HarperPerennial, 2007.

Engel, Michael S. *Innumerable Insects: The Story of the Most Diverse and Myriad Animals on Earth*. New York: Sterling, 2018.

Goulson, Dave. *Silent Earth: Averting the Insect Apocalypse*. New York: Harper, 2021.

Haskell, David George. *Sounds Wild and Broken: Sonic Marvels, Evolution's Creativity, and the Crisis of Sensory Extinction*. New York: Viking, 2022.

Heinrich, Bernd. *Life Everlasting: The Animal Way of Death*. Boston: Mariner Books, 2013.

Hirshfield, Jane. *Ledger*. New York: Knopf, 2020.

Kimmerer, Robin Wall. *Braiding Sweetgrass: Indigenous Wisdom, Scientific Knowledge, and the Teachings of Plants*. Minneapolis: Milkweed Editions, 2013.

Kolbert, Elizabeth. *The Sixth Extinction: An Unnatural History*. New York: Henry Holt, 2014.

Lanham, J. Drew. "A Convergent Imagining: What If Martin Luther King Jr. and Rachel Carson Had Met?" Literary Hub, February 22, 2021. https://lithub.com/a-convergent-imagining-what-if-martin-luther-king-jr-and-rachel-carson-had-met/.

McFarlane, Robert, and Jackie Morris. *The Lost Words*. Toronto: Anansi International, 2018.

Nezhukumatathil, Aimee. *World of Wonders: In Praise of Fireflies, Whale Sharks, and Other Astonishments*. Minneapolis: Milkweed Editions, 2020.

Oliver, Mary. *New and Selected Poems*. Boston: Beacon Press, 1992.

Raffles, Hugh. *Insectopedia*. New York: Pantheon, 2010.

Sheldrake, Merlin. *Entangled Life: How Fungi Make Our Worlds, Change Our Minds and Shape Our Futures*. New York: Random House, 2020.

Simard, Suzanne. *Finding the Mother Tree: Discovering the Wisdom of the Forest*. New York: Knopf, 2021.

Solnit, Rebecca. *Hope in the Dark: Untold Histories, Wild Possibilities*. Chicago: Haymarket Books, 2016.

Tallamy, Douglas. *Nature's Best Hope: A New Approach to Conservation That Starts in Your Yard*. Portland, OR: Timber Press, 2020.

Van Horn, Gavin, Robin Wall Kimmerer, and John Hausdoerffer, eds. *The Kinship Series: A World of Relations*. Libertyville, IL: Center for Humans and Nature, 2021.

Wilson, E. O. *Half-Earth: Our Planet's Fight for Life*. New York: Liveright, 2013.

Wilson, Joseph S., and Olivia Messinger Carril. *The Bees in Your Backyard: A Guide to North America's Bees.* Princeton, NJ: Princeton University Press, 2016.

Xerces Society. *100 Plants to Feed the Bees: Provide a Healthy Habitat to Help Pollinators Thrive*. North Adams, MA: Storey, 2016.

Playlist

Where the Grass Still Sings playlist. Compiled by Joe Parisi, playlist master. https://open.spotify.com/playlist/6KLL3g4bAJmKRIZEqNUnFd?si=muE_ljZTTa68Gp7HPQtoHw.

Credits

An earlier version of chapter 1 was published in *Edge Effects*, March 13, 2018.

An earlier version of chapter 2 was published in *Catapult*, April 8, 2020.

An earlier version of chapter 3 was published in *Emergence Magazine*, May 27, 2021, with photos by Felipe Villegas.

An earlier version of chapter 4 was published in *Edge Effects*, February 8, 2022.

An earlier version of "Claire Morgan: Dead Owls and Bluebottle Flies" was published in *The Learned Pig*, October 18, 2020.

An earlier version of chapter 8 was published in *Belt Magazine*, October 19, 2018.

An earlier version of the afterword was published in *About Place Journal* 4, no. 2 (2016).

Wendell Berry, "The Peace of Wild Things," from *New Collected Poems*. Copyright © 2012 by Wendell Berry. Reprinted with the permission of The Permissions Company, LLC on behalf of Counterpoint Press, counter pointpress.com.

Nickole Brown, "Magicicada Septendecim," was reprinted with permission of the poet.

Brenda Cárdenas, "Lampyridae," from *Trace* (Pasadena, CA: Red Hen Press, 2023).

James Crews, "Monarch," was reprinted from *Bluebird*, published by Green Writers Press in 2020, with permission of the poet.

Catherine Jagoe, "Cherry-Faced Meadowhawk," was reprinted with permission of the poet.

Mary Oliver, "The Summer Day." Reprinted by the permission of The Charlotte Sheedy Literary Agency as agent for the author. Copyright © 1990, 2006, 2008, 2017 by Mary Oliver with permission of Bill Reichblum.

Heather Swan, "On the Day After You Left this World," was first published by *One Art*, March 7, 2021.

Art by Jennifer Angus reproduced by permission of the artist.

Art by Emily Arthur reproduced by permission of the artist.

Art by Lea Bradovich reproduced by permission of the artist.

Art by Susan Carlson reproduced by permission of the artist. Copyright Susan Carlson (susancarlson.com).

Art by Liz Anna Kozik reproduced by permission of the artist. Website: liz.kozik.net. Email: liz@kozik.net.

Sculptures by Edouard Martinet reproduced by permission of the artist. Website: edouardmartinet.fr. Instagram: @ edouardmartinet.

Art by Rosalind Monks reproduced by permission of the artist.

Art by Claire Morgan © Claire Morgan. Courtesy Galerie Karsten Greve (Köln, Paris, St. Moritz).

Art by Amy Spassov reproduced by permission of the artist.